The Physics of Theism

The Physics of Theism

God, Physics, and the Philosophy of Science

Jeffrey Koperski

WILEY Blackwell

This edition first published 2015
© 2015 John Wiley & Sons, Ltd.

Registered Office
John Wiley & Sons, Ltd, The Atrium, Southern Gate, Chichester, West Sussex, PO19 8SQ, UK

Editorial Offices
350 Main Street, Malden, MA 02148-5020, USA
9600 Garsington Road, Oxford, OX4 2DQ, UK
The Atrium, Southern Gate, Chichester, West Sussex, PO19 8SQ, UK

For details of our global editorial offices, for customer services, and for information about
how to apply for permission to reuse the copyright material in this book please see our
website at www.wiley.com/wiley-blackwell.

The right of Jeffrey Koperski to be identified as the author of this work has been asserted in
accordance with the UK Copyright, Designs and Patents Act 1988.

Library of Congress Cataloging-in-Publication Data

Koperski, Jeffrey.
    The physics of theism : god, physics, and the philosophy of science / Jeffrey Koperski.
        pages   cm
    Includes bibliographical references and index.
    ISBN 978-1-118-93281-0 (cloth) – ISBN 978-1-118-93280-3 (pbk.)
    1. Physics–Religious aspects.   2. Theism.   3. Religion and science.
    4. Science–Philosophy.   I. Title.
    BL265.P4K67 2015
    211′.3–dc23
                                                                    2014025620

A catalogue record for this book is available from the British Library.

Cover image: © Brett Charlton / iStockphoto

Set in 10.5/13pt Minion by SPi Publisher Services, Pondicherry, India

Printed in Singapore by C.O.S. Printers Pte Ltd

1   2015

To my family, Marie, Andrew, Marcus, and Christopher
without whom this book
would have been completed twelve months sooner.

Not a bad tradeoff.

Contents

Introduction

I.1 Maps

History remembers the names of famous explorers, and it's easy to see why. Discovery is full of intrigue. It sparks the imagination. Once the explorers have returned home, another group comes along: the cartographers. They are somewhat less well known, and that also makes sense. The blazing of the trail is a good story; map making … not so much. Still, if new territory is to be accessible, what most of us need is a map of the terrain.

Something analogous is true in every academic discipline. There are those on the cutting edge, breaking new ground and offering fresh insights. Then, there are those who sort it all out, mapping out the camps and explaining how things stand. Although I've done a bit of exploring, by nature I am a cartographer. Explorers' notes are often messy and hard to understand. This book is a map of an unfamiliar terrain and a guide through it. The territory of interest is found along the border of science and religion.

Theologians and philosophers of religion often look to science, especially physics, for ideas. They want to know how the world was created, how God might interact with it, and whether there are any fingerprints of divine action. In the chapters that follow, we will consider how physics is relevant to matters of religion, and more surprisingly perhaps, how the influence sometimes goes the other way. If you want to know what quantum mechanics, relativity, and chaos theory have to do with religious belief, this is a good place to start.

Very well, but why then is a philosopher writing this book? If we're talking about science and what it means, we usually hear from a physicist.

The Physics of Theism: God, Physics, and the Philosophy of Science,
First Edition. Jeffrey Koperski.
© 2015 John Wiley & Sons, Ltd. Published 2015 by John Wiley & Sons, Ltd.

You never see a philosopher on CNN addressing these questions. You never see a philosopher on CNN addressing *any* questions.

While that's true, people outside of the ivory tower often don't understand the hyperspecialization of academia these days. Few scholars are able to keep up with trends even in their own discipline. In addition, one's training and expertise are suited to specific needs, especially in science. Experimentalists are experts in the collection and analysis of data. Theoreticians are in the best position to develop and judge between competing theories. This division of labor means that no one is simply an expert on "physics," let alone the whole of science. Still, it's hard not to cross disciplinary borders on occasion. Scientists sometimes offer opinions on matters of religion, although only their negative remarks generally make the news. And insofar as the truth about physical reality is relevant, scholars in religion and the humanities want to be informed. This explains the proliferation of conferences, workshops, and centers devoted to the study of science and religion.

While these conferees might not realize it, the terrain on which this discussion takes place is most often the philosophy of science. Within the humanities, philosophers of science generally pay the closest attention to the goings-on in science as well as its history. They make generalizations about the nature of scientific inferences, the assumptions and implications of science, and how each of these has changed over time. In short, philosophers of science specialize in just the sort of questions that tend to emerge in the science-and-religion literature. Hence, we tend to make good guides for this terrain (or at least that's what I've talked my publisher into believing). This book aims at mediating a wide range of debates in which science, especially physics, plays a significant role in matters of religion.

The reader should know that while I try to give an accurate description of the issues to be discussed, this is not a "neutral" textbook that one might use in introductory courses. Like any philosopher, I have views on these matters, and there are judgment calls to make at every turn. Not everyone will agree with my analysis (but they should). To see what this approach looks like, let's briefly consider the recent history of cosmology.

I.2   Cosmology: Singularity and Creation

Modern cosmology has never just been about science. Although Einstein's field equations for general relativity showed that the universe would expand or contract over time, that idea did not square with his philosophical views.

Einstein believed instead that the whole of space was "static." To bring the physics in line with what his philosophy said it should be, Einstein added the infamous cosmological constant to his equations, a move he would deeply regret.

The first widely accepted solutions of Einstein's field equations predicted that our universe has not always existed. (More precisely, the Friedmann–Lemaître–Robertson–Walker (FLRW) models have a finite time metric.)[1] Many theists, including Pope Pius XII, were delighted by what came to be known as the Big Bang since it seemed to confirm something like creation *ex nihilo*. As astronomer Robert Jastrow put it,

> For the scientist who has lived by his faith in the power of reason, the story ends like a bad dream. He has scaled the mountains of ignorance; he is about to conquer the highest peak; as he pulls himself over the final rock, he is greeted by a band of theologians who have been sitting there for centuries. (1992, 107)

Although Jastrow was an agnostic, this quote has been used by theists ever since its publication.

As one might imagine, atheists reacted differently.[2] Many, like physicist William Bonnor, took the Big Bang as a religious doctrine masquerading as science:

> The underlying motive [for Big Bang cosmology] is, of course, to bring in God as creator. It seems like the opportunity Christian theology has been waiting for ever since science began to depose religion from the minds of rational men in the 17th century. (Kragh 2004, 241–242)

Astronomer Fred Hoyle actually coined the term "Big Bang" as a pejorative, declaring it "a form of religious fundamentalism" (Kragh 2004, 235). All this motivated a search for solutions that did not entail a finite beginning. The most successful of these was the steady-state model in which the universe was infinitely old and matter continually created throughout space, not just once at the Big Bang. The steady-state model was seen by many on both sides as being antitheistic or at least undercutting the need for a creator, as Carl Sagan argued: "This is one conceivable finding of science that could disprove a Creator—because an infinitely old universe would never have been created" (Halvorson and Kragh 2011, sec. 3). The debate between the two rival views ranged from whether one was more scientific than the other to questioning the scientific status of cosmology itself.[3]

While the steady-state model was abandoned in the mid-1960s,[4] the search for alternative cosmologies goes on and religious beliefs continue to play a role. One unsolved question is whether the Big Bang had a cause. The universe exists, but *why* does it exist? Why is there a universe—galaxies, quasars, and the rest—rather than nothing at all? Cosmologist and self-described "antitheist" Lawrence Krauss purports to give an answer in his recent book *A Universe from Nothing: Why There Is Something Rather Than Nothing*. It contains a "scientific"—that is, nonphilosophical and nontheological—explanation for why there is a universe. Note, the question is not merely why does this universe exist, but why is there anything at all. Many have argued that the answer must be something outside the cosmos, what Aristotle called the First Cause and what most theists call God. As the title of his book declares, Krauss's view is that the universe need not have been created. It sprang up from nothing. One motivation for this, it would seem, is the undermining of theism. Richard Dawkins sums it up this way in his afterword to the book:

> Even the last remaining trump card of the theologian, "Why is there something rather than nothing?," shrivels up before your eyes as you read these pages. If *On the Origin of Species* was biology's deadliest blow to supernaturalism, we may come to see *A Universe From Nothing* as the equivalent from cosmology. The title means exactly what it says. And what it says is devastating. (Krauss 2012a, 191)

Claiming to have solved a longstanding metaphysical question, Krauss's arguments got the attention of philosophers. While he has a lot of interesting things to say, the philosophers were, well, unimpressed. The issue has to do with what exactly the physics entails. Let's grant that everything Krauss says about the science is correct. Has physics, even highly speculative physics, shown that the universe could have spontaneously come into existence from nothing? As philosopher David Albert (2012) points out, Krauss's "nothing" is somewhat peculiar. Among other things, it changes according to the laws of quantum field theory. But wait: how did quantum mechanics get in here? I thought we were talking about nothing. It turns out that Krauss's "nothing" is somewhat of a misnomer. His version of nothing has physical properties and contains relativistic quantum fields. Albert and others question whether such a well-defined physical entity counts as nothing. Krauss has since backpedalled a bit and claims that he doesn't really care what philosophers and theologians mean by the word (Krauss

2012b). (As fellow cosmologist Sean Carroll notes, Krauss doesn't have to care, "but if the subtitle of your book is 'Why There Is Something Rather Than Nothing,' you pretty much forfeit the right to claim you don't ..." (Carroll 2012).)

For our purposes, all this serves as a nice illustration of the interplay between science, philosophy, and religious belief. What Krauss has to say is interesting and important. His arguments should be carefully considered by theologians and philosophers. Krauss himself has been pushed to be more accurate and precise about his claims. The back and forth between scholars of different disciplines pares away overstatements from real advances in scientific knowledge. It also helps make clear what the implications of physics are for matters of philosophy and religion. It is this sort of interdisciplinary crossover that we will have an eye on throughout the rest of this book.

I.3   Overview

We are not finished with cosmology; but for now, let's briefly consider what is to come.

I.3.1   *Science and Religion: Some Preliminaries*

Skeptics often claim that science and religion are in conflict. Others say that the two realms are too different for there to even be a conflict. As we will see in Chapter 1, neither of these is the best way to understand the relation between science and religion. To understand why, we need a clearer picture about the nature of science itself. To do that, we begin with its history. As it turns out, the conventional wisdom about science and religion is deeply flawed. The relationship between the two is more subtle and complex than is usually assumed. One reason for this is the role of *metatheoretic shaping principles*. Such principles capture scientists' views about the nature of the physical world and how best to study it. If you've never heard of shaping principles, that's because they are rarely noticed. We generally think of them as "just the way things are" from a scientific viewpoint. Shifts in these principles are only evident across broad stretches of history. As we will see, religious beliefs have had a surprising role in their development since the beginning of the scientific revolution.

I.3.2   Fine-Tuning and Cosmology

One of the standard topics in any Introduction to Philosophy course is the teleological argument for the existence of God, more commonly known as "the argument from design." Versions of this argument can be found in ancient times down through Paley's famous watch analogy. As we will see in Chapter 2, things got more interesting about 30 years ago. It turns out that the universe is a bit like an aquarium. For life to be possible, two dozen or so cosmological variables must have values within extremely narrow ranges. Change any one by even a slight amount and living creatures could not exist here or anywhere else in the universe. That is not what physicists expected. The universe shouldn't care whether life exists or not. Why then do so many of its fundamental parameters seem to be set to the precise values needed for our existence? Most physicists and philosophers believe that fine-tuning needs an explanation. Theism, of course, provides one answer: The universe looks fine-tuned for life because it has been fine-tuned for life. Our cosmic environment bears the earmarks of design. In this chapter, we consider some examples of fine-tuning, the best naturalistic explanations for it, and whether the need for explanation is itself based on faulty premises.

I.3.3   Relativity, Time, and Free Will

Chapter 3 presses into an old concern for philosophers: free will. Physics has played a significant role in the conversation, often by undermining the possibility of freedom. Some varieties of determinism were grounded in Newtonian mechanics: If the behavior of all things, including the atoms in our own bodies, is wholly determined by the laws of physics, then there doesn't appear to be any room left for free will. In such a world, a kicker doesn't choose to kick a field goal any more than the football chooses to go through the goal posts. It's all just a matter of the laws of physics working themselves out.

No one worries about that particular form of determinism now that Newtonian physics has been replaced by quantum mechanics. The story, however, does not end there. Einstein's theory of relativity also undermines free will as well as our commonsense view of time. According to the most straightforward reading of relativity, time does not flow, and there is no real difference between what we think of as the past and future. From the four-dimensional perspective demanded by relativity, almost all of our beliefs about time are based on an illusion. The future—or at least what we already think of as the future—exists, and nothing that happens in the present can change it. (Why didn't anyone mention that in my freshman physics class?) In Chapter 3, we will consider some ways of reestablishing a flow of time within the "block universe" of relativity, the unique reality of the present, and where to find room for free choice.

I.3.4   Divine Action and the Laws of Nature

Ever since the notion of a "law of nature" took hold in science, philosophers and theologians have questioned God's relationship to those laws. In some theological circles today, it is taken for granted that God would seldom, if ever, violate laws that God himself has ordained. At the same time, most theists believe that God answers prayers and at times acts within the natural order. How then can God act without violating his own laws? Since quantum mechanics is not deterministic, many theologians and theistic scientists believe that God works within the random gaps of quantum indeterminacy. The accumulation of such small changes, they argue, can produce macroscopic effects. In Chapter 4, we will consider what would it be for God to violate the laws of nature and what range of activity is possible by noninterventionist means. I will argue that much of this debate should be reconsidered on both scientific and theological grounds.

I.3.5   Naturalisms and Design

Intelligent design (ID) has been a controversial topic over the past decade. While ID is usually associated with evolution, the relation between design arguments and naturalism transcends biology. How one answers the questions involving ID ramifies across the other sciences, including physics. While much has been written, it seems that even scholars cannot help but get caught up in the culture war aspects of the controversy. In Chapter 5, I attempt to reorient the debate away from *ad hominem* attacks and questions about motives. Philosophers of science have made important advances in our understanding of anomalies, theory change, and background beliefs over the last 40 years. The ID debate can benefit from this work. Philosopher Larry Laudan's analysis of young earth creationism in the 1980s serves as an important model.

I.3.6   Reduction and Emergence

Reductionism is the view that, in principle, high-level theories, laws, and complex entities can be explained by or reduced to a more basic level: Psychology can be reduced to neurophysiology, neurophysiology to molecular biology, molecular biology to organic chemistry, all the way down to quantum field theory. While reductionism is not wholly a matter of physical science, physics plays a key part since it is thought to describe the ground floor of reality. Among analytic philosophers, this form of reductionism is often considered to be a failed project. Theists have been keen on this development since, of course, God cannot be reduced to physics. Many philosophers and philosophically informed scientists are turning to the notion of *emergence* as an alternative to reduction. Might the mind, for example, be an autonomous entity that emerges from but is not identical to the brain? Might each of the levels of reality above fundamental physics be irreducible and emergent? What exactly does that mean? We will consider these questions and assess this new emphasis on emergence mostly by using examples within physics itself.

I.3.7   The Philosophy of Science Tool Chest

Chapter 7 contains some suggestions for how tools in the philosophy of science can help scholars in religion, theology, and the philosophy of religion. These include matters of theory choice, anomalies and theory change, truth and approximate truth, underdetermination, and realism/antirealism. As esoteric as those might sound, they are useful for understanding a number of questions including the nature of religious belief, the relationship between religious traditions, and the role of faith.

Before we begin, I would like to acknowledge those colleagues and friends who helped make this project a success. Helpful comments on the text were provided by Chris Arledge, Peter Brian Barry, Ron Benson, Robert Bishop, Robin Collins, Tammy DeRuyter, Hans Halvorson, Lorna Holmes, Aaron Kostko, Al Lent, Alan Love, Bradley Monton, Bob O'Connor, Brian Pitts, John Polkinghorne, Del Ratzsch, David Raup, Andrés Ruiz, Bob Russell, David Schubert, Walter Schultz, Charles Taliaferro, Paul Teed, and Dale Tuggy. A very special thanks to Philip West, who read the entire manuscript, offering helpful advice along the way. Finally, thanks to Rodney Holder and Thomas Tracy who reviewed the book for Wiley-Blackwell and provided very helpful comments and corrections.

Three chapters are based in part on previously published articles, all used by permission: chapter 2, "Should We Care about Fine-Tuning?" *British Journal for the Philosophy of Science* 56 (2005): 303–319; chapter 4, "God, Chaos, and the Quantum Dice," *Zygon* 35 (3) (2000): 545–559; chapter 5, "Two Bad Ways to Attack Intelligent Design and Two Good Ones," *Zygon* 43 (2) (2008): 433–449. Two grants also supported parts of this work: *Randomness and Divine Action* (Calvin College, chaps 4, 6, 7) and *The Emergence of Biological Complexity* (Cambridge-Templeton Consortium, chap. 6).[5] Thanks also to Saginaw Valley State University, my home institution, for a faculty research grant and sabbatical leave.

Each chapter is mostly a standalone piece, although there are occasional references made to material found elsewhere in the book. When that happens, there is a citation for the appropriate chapter and section. Of course, reading every word from cover to cover would be most beneficial to the reader and humanity at large; but if you would rather jump around a bit, you should be able to do so and still understand what's going on.

Notes

1 Which is not the same as the universe having an identifiable "first moment," as Halvorson and Kragh emphasize (2013, 244). The mathematical limits involved ought not be thought of as simply counting backward in time to an instant of creation.
2 Many but not all atheists, as Kragh (2004, 242) notes. There were atheists who supported Big Bang cosmology and theists who opposed it.
3 For some of the details, see Kragh (2004, 233–242).
4 This was largely due to the accidental discovery of cosmic microwave background radiation by American radio astronomers Arno Penzias and Robert Wilson. This radiation is a leftover of the Big Bang and could not be accounted for by the steady-state model.
5 Both of which were made possible by grants from the John Templeton Foundation, whose lawyers would like me to remind you that the views expressed here are mine, not necessarily theirs.

References

Albert, David Z. 2012. "On the Origin of Everything." *The New York Times*, March 23, sec. Sunday Book Review. http://www.nytimes.com/2012/03/25/books/review/a-universe-from-nothing-by-lawrence-m-krauss.html. Accessed June 6, 2013.

Carroll, Sean. 2012. "A Universe from Nothing?" *Cosmic Variance*. http://blogs.dis covermagazine.com/cosmicvariance/2012/04/28/a-universe-from-nothing/. Accessed June 6, 2013.

Halvorson, Hans, and Helge Kragh. 2011. "Cosmology and Theology." In *Stanford Encyclopedia of Philosophy*, edited by Edward N. Zalta. Stanford: Metaphysics Research Lab, Center for the Study of Language and Information, Stanford University. http://plato.stanford.edu/entries/cosmology-theology/. Accessed August 2, 2013.

Halvorson, Hans, and Helge Kragh. 2013. "Theism and Physical Cosmology." In *The Routledge Companion to Theism*, edited by Stewart Goetz, Victoria Harrison, and Charles Taliaferro, 241–255. New York: Routledge.

Jastrow, Robert. 1992. *God and the Astronomers*. New York: W.W. Norton.

Kragh, Helge. 2004. *Matter and Spirit in the Universe: Scientific and Religious Preludes to Modern Cosmology*. London: Imperial College Press.

Krauss, Lawrence M. 2012a. *A Universe from Nothing: Why There Is Something Rather Than Nothing*. New York: Free Press.

Krauss, Lawrence M. 2012b. "The Consolation of Philosophy." *Scientific American*, April 27. http://www.scientificamerican.com/article.cfm?id=the-consolation-of-philos. Accessed June 6, 2013.

1

Science and Religion: Some Preliminaries

1.1 Conventional Wisdom

Science and religion have been at war with one another since Galileo was tortured by the Inquisition.

The Catholic Church taught that the earth was flat until Christopher Columbus proved otherwise.

The scientific revolution finally freed Europe from the grip of religion.

As every historian of science knows, these three nuggets of conventional wisdom are false. Galileo was never jailed, let alone tortured. Aristotle knew the Earth was round and so did nearly every educated person in the Middle Ages.[1] The "war" between science and religion? That was a rhetorical invention of the 19th century. As we'll see, most of the key figures in and around the scientific revolution believed that philosophy, theology, and science were compatible if not complementary disciplines. Some, like Descartes and Pascal, made contributions in all three.

The intellectual landscape is now very different, of course. The pursuit of knowledge is now so highly specialized that practitioners have a hard time communicating with others in their own field let alone those in other disciplines. Academics are therefore cautious about straying too far from their area of expertise. That is until we turn to the topic of religion. Then everyone has an opinion. The same goes for science in general rather than, say, solid-state physics or tropical entomology. Everyone seems to know what

The Physics of Theism: God, Physics, and the Philosophy of Science,
First Edition. Jeffrey Koperski.
© 2015 John Wiley & Sons, Ltd. Published 2015 by John Wiley & Sons, Ltd.

science is and is not. "It's all quite simple actually. Science is based on reason and empirical investigation. Religion is based on faith. Next question." Philosophers and historians of science have long recognized that things aren't that tidy. Physics and metaphysics were not always studied by different departments within the university, and the modern view of religion as a private, spiritual matter was not always the norm.

To understand the relation between science and religion, we begin this chapter with some history. It should be no surprise that we start in ancient Greece, tracing the influence of Aristotelian thought into the late Middle Ages. A turning point occurs in the 14th century with attacks on Aristotelian/Thomism. This shift reverberates through Galileo, Descartes, Boyle, and the early modern era. After the overview of history, we will consider the overall structure of science and several models used to describe its relationship to religion. Getting a handle on this will prove to be more difficult than one might think. At the end, I will argue that there is no single model that can capture the complex relation between science and religion. The best we can hope for is broad themes that show how the two fields influence one another.

1.2   History

1.2.1   Ancient Greece

While Plato and Aristotle were not the first important thinkers to come from ancient Greece, they were the most influential. Like Pythagoras and Parmenides before him, Plato believed that the things with the greatest degree of reality were not what we can touch and see. The visible world is but a pale and imperfect copy of ultimate reality, which is invisible, immaterial, and timeless. What is most real for Plato resides in the perfect and eternal realm of the Forms. While we see particular instances of triangles, justice, and beauty, they are imperfect reflections of the pure Forms of Triangularity, Justice, and Beauty. Platonic knowledge consists in understanding the Forms themselves. The principal task of the philosopher, Plato taught, is to get beyond our own sensory experiences to the truth of the Forms. His famous "Allegory of the Cave" (Book VII of *The Republic*) portrays the struggle to put aside how things seem and push on to the true nature of reality.

While the roots of science run through Plato, he is less important for our purposes than his best known student: Aristotle. Early modern figures like

Descartes and Galileo were reacting against an Aristotelian philosophy which had reached its peak in the 13th century. While Aristotle rejected the far-off reality of the Forms, he did not believe that the true nature of the world was obvious to the average person. Consider a horse. What makes that entity a horse rather than an oak tree or ruby? And why is it that all horses share certain traits? According to Aristotle, the horse—like everything else—is composed of two things: prime matter and an essence (or *substantial form*). A particular horse is the "hylomorphic composition" of the distinctive essence of a horse with matter. This essence is the collection of properties (or *universals*) that make a substance what it is. The essence is also what gives an entity its capacity to act, whether living or nonliving, animate or inanimate. For example, if you pick up a stone and release it, it falls to the ground. Why? Gravity of course, but that idea would not be developed for another 2000 years. For Aristotle, solid objects naturally move toward the center of the Earth, then thought to be the center of the universe as well. It is part of their essence to do so. Likewise, fire naturally tries to reach up to the celestial realm. Nothing makes fire behave that way; that's simply what it does by nature, again, in accordance with its essence. Horizontal motion is contrary to the nature of solid objects, a fact that Aristotle thought he could prove. Put a book on the desk. Now, push it sidewise. It stops after you take your fingers away. Why? Because the internal goal of the book, its "final cause" in Aristotelian terms, is to get to the center of the Earth. It doesn't want to move horizontally. "Violent motions" like horizontal displacement can be imposed on objects, but it isn't what they do by nature.

In the Aristotelian view developed by Aquinas and other medieval philosophers, understanding physical reality meant discovering the underlying substantial form of each thing. "Natural philosophy"—what later came to be known as science—centered on the discovery and study of the universals comprising each essence.[2]

Studying essences mostly required "insight" (*epagōgē*) rather than a lot of careful observation. On the Aristotelian/Thomist view, the senses merely provide raw data for the intellect, where reasoning takes place. We have direct access to particulars: these horses, those trees, etc. By contemplating these observations, the intellect is able to abstract the universals common to a set of particulars arriving at, as Aquinas says, an "understanding of the very substance of that being" (*Summa Contra Gentiles*, 1.3.3). Consider a simple example: the essence of a triangle. After examining several triangles and contrasting them with other figures of plane geometry, it's clear that three-sidedness is part of the essence of every triangle. There are no worries that some mathematician on an alien world might have discovered triangles without three sides. Note that this is not a matter of defining the word 'triangle.' Aristotelian universals are things we discover; they are already out there to be known. We do not invent them by fixing a definition.

The upshot of this is that experimentation was not highly valued. Once one had grasped the essence of a thing, Aristotelians saw no need for further investigation. They therefore did not generally test their ideas about substantial forms. Artificial experiments were thought to produce violent behavior—counter to an object's nature—rather than the natural behavior determined by essential properties. Once the works of Aristotle were rediscovered in 12th-century Europe, they became standard texts. Science at the universities often meant studying Aristotle and his commentators, not empirical investigation (McMullin 1967, 335–337). (There is some irony in this as Aristotle himself did a great deal of empirical study, especially in biology.)

Although Plato's metaphysics is usually contrasted with Aristotle's, both held that nature is governed by timeless, unchanging principles. There is no sense in which the Forms or essences could have been something other than they are. The ground floor of reality has no contingency; it does not depend on anything else for its existence and could not be other than what it actually is. The foundation, whatever the precise details, is timeless, fixed, and necessary. As philosopher Del Ratzsch stresses,

> nearly all the Greek philosophers believed that on its most fundamental level, ultimate reality—whether that was matter or atoms or immaterial principles or Forms—was eternal, fixed, unchanging, and governed by structures and principles of reason.
>
> Given this rigid, logical structuring of the ultimate, governing level of reality, most Greeks thought that any 'nature' or 'world soul'—and even the gods themselves—were subject to, and had to work within or around, the boundaries imposed by this eternal, rigid, ultimate order of reality. (2010, 56)

As one might imagine, this became a theological issue for Christians, Jews, and Muslims (Christians in particular, once the Bishop of Paris condemned 219 Aristotelian propositions as heretical in 1277). On one hand, God is omnipotent. On the other hand, not even God can change the essence of a thing. If one takes a triangle and removes the essential property of three-sidedness, one no longer has a triangle. Moreover, God himself has a nature,

which includes omniscience, goodness, and absolute rationality. God therefore is not free to act in any way he might choose. God is limited to those
choices that the divine nature would permit.[3]

Not everyone was happy with this conclusion.

1.2.2   *Voluntarism and Nominalism*

One reaction in the 14th century was the rise of *voluntarism*: in short, God
can choose to do whatever he wants. He is not restricted by his essence.
Consider ethics and God's commands. Thomists held that God commands
what he does because he is perfectly good and rational. Voluntarist, like
William Ockham, argued instead that what God commands is primarily a
matter of will. He simply chooses to require certain actions and forbid
others. Both agree that God would never lie, but for Thomists this is because
God is omniscient, rational, and good. God knows that lying is wrong and
therefore will not do it. In contrast, voluntarists believe that God does not
lie simply because he has chosen not to.

Philosophers at the time also began rethinking the received view of
substantial forms. One worry was that they are "occult entities": we can't
see them. They can only be discovered by abstraction. More importantly,
appealing to hidden essences in order to explain observable traits began
to be seen as hopelessly obscure. Any action or property could be "explained"
simply by declaring it to be the product of an essential property. A famous
example is from Molière's *Tartuffe* where a doctor explains why opium
makes one sleepy. It is because, he says, of its "virtus dormitiva"—the
essential capacity to induce sleep. Perhaps, said the critics, but in what
sense does that explain anything? A rival metaphysical view known as
nominalism emerged in response. Nominalists like Ockham and Peter
Abelard argued that essences (and universals in general) are merely concepts in the mind. They aren't "out there" as independent parts of reality
to be discovered. Property terms like 'red' and 'triangular' do not refer
to abstract entities. Whatever commonality exists among red objects is
merely a matter of perception. For nominalists, grasping a universal was
a purely mental exercise rather than the acquiring of deep insight into
nature. Real knowledge was limited to the behavior of particulars.
However, that sort of knowledge depended on observation rather than
pure reason and was therefore considered less certain. While one can
observe regularities in nature, such generalizations were thought to be
fallible and approximate (McMullin 1967, 339–340).

In addition to being obscure, many began to see substantial forms as useless intermediaries that God did not need in governing the universe. An omnipotent, omniscient being does not require essences embedded in prime matter in order to get things to work the way he wants. According to Robert Boyle, the medieval view undermines

> the honor of the great author and governor of the world, that men should ascribe most of the admirable things, that are to be met with in it, not to him, but to a certain nature. … For my part, I see no need to acknowledge any architectonic being besides God. … Those things which the [medieval] school philosophers ascribe to the agency of nature interposing according to emergencies, I ascribe to the wisdom of God. (Deason 1986, 180–181)

If substantial forms do all the work, then the creative activity of God becomes less apparent. And since God, the omniscient architect, doesn't need such entities anyway, many like Boyle began to look for alternatives.

1.2.3 *Mechanistic Philosophy*

The medieval picture of nature had been organic. The idea that rocks strive to return to the center of the Earth was the same as trees attempting to send roots to sources of water. "Strive" and "attempt" are not metaphors on this view. Every being was thought to have a purpose or end (*telos*) determined by its essence. When this nature-as-organism view fell out of favor, a new one arose: nature-as-machine. On the new view, matter was no longer considered active and lifelike, but passive and inert. Change does not arise from within an entity according to the new mechanistic philosophy, but from external forces. Like the gears and springs in a clock, causation is limited to one body pushing or pulling on another.

The machine analogy soon dominated the study of nature. Early modern scientists believed the universe was itself an artifact created by a rational, intelligent agent and that we have been given minds by God in order to understand how it works. With sufficient study, we should be able to determine the principles God used, as Kepler makes clear,

> God, who founded everything in the world according to the norm of quantity, also has endowed man with a mind which can comprehend these norms. (1597 letter to Maestlin)

Those [norms] are within the grasp of the human mind. God wanted us to recognize them by creating us after his own image so that we could share in his own thoughts … and, if piety allows us to say so, our understanding is in this respect of the same kind as the divine, at least as far as we are able to grasp something of it in our mortal life. (1599 letter to Johannes Georg Herwart von Hohenburg)

Two centuries later, geologist James Hutton would begin his *Theory of the Earth* with similar thoughts:

When we trace the parts of which this terrestrial system is composed, and when we view the general connection of those several parts, the whole presents a machine of a peculiar construction by which it is adapted to a certain end. We perceive a fabric, erected in wisdom, to obtain a purpose worthy of the power that is apparent in the production of it. ([1788] 2007, 11)

None of this entailed that humans could understand God's design principles with absolute accuracy and precision, however. Galileo argued that our intellectual resources are limited; God's are not. Hence, what we believe to be the norms and principles God used might only be good approximations (Davis 1999, 82).

All this leaves open the question about what these norms and principles are. The answer is given in the preface to the 1st edition of Newton's *Principia Mathematica*: "[The] moderns, rejecting substantial forms and occult qualities, have endeavored to subject the phenomena of nature to the laws of mathematics…" ([1687] 1966, 1, xvii). Just as kings proclaimed the laws for a country, God was thought to decree laws for nature, and the language he chose was mathematics. This was not merely a loose way of speaking, as historian John Hedley Brooke points out. "When natural philosophers referred to *laws* of nature, they were not glibly choosing that metaphor. Laws were the result of legislation by an intelligent deity" (1991, 19).

The break from Aristotelianism was now complete. For ancient Greeks, *nomos* had to do with law and convention in contrast to *phusis*, the realm of nature. Laws were contingent on human will; nature was mind independent. So while the phrase *nomos phusis* can occasionally be found in ancient sources,[4] the idea that nature itself was law governed fully emerged from a theistic worldview coupled with the new mechanical philosophy.[5]

Laws such as the conservation of momentum and *vis viva* (what would later be called kinetic energy) had an especially close tie to theology.

Consider Descartes' three laws of motion. The first is that bodies only change state "as a result of external causes." The second is that without such influence bodies will only move in straight lines. According to Descartes, "The reason for [the second] rule is the same as the reason for the first rule, namely the immutability and simplicity of the operation by which God preserves matter in motion" (*Principles of Philosophy* §2.39). Descartes' laws of motion became the starting point for others such as Wallis, Huygens, and Newton.

By the seventeenth century, voluntarism and nominalism had for the most part won the debate, especially in England.[6] Most early scientists in Europe believed that the universe is an artifact designed by God, who is the only necessary and eternal entity, and that the whole of nature is subject to God's direct will (i.e., no Aristotelian intermediaries). Everything in nature is created and contingent on God's free choices, including the laws of nature. Boyle sums up much of this in an unpublished paper quoted by historian Ted Davis (the archaic spelling and style in the original have been updated here):

> [Since] then there was no being besides God himself who is eternal, and beings … that had a beginning must derive their natures and all their faculties from his arbitrary will; and consequently man himself and all intellectual as well as all corporeal creatures were but just such as he thought fit to make them; and as he freely established … the laws of motion by which the universe was framed and doth act; so he freely constituted the reason of man and other created intellects and gave them those ideas and measures of truth by which they are guided in all their ratiocinations. And the very axioms or most acknowledged truths being but relations resulting from the nature of the mind …, God might have so ordered things that propositions very differing from these might have been true. As he might have contrived the palate, that honey had tasted bitter to men, and gall sweet. (Davis 1999, 86)

There are three main claims here. First, God has determined the nature of every creature and the laws of nature themselves. Second, God has created human intellect with the ability to discover truth about such things. Third, God might have freely chosen to do it all some other way. Both nature itself and our means for discerning it could have been completely different.

While opinions varied about the range of God's free choices, Descartes' voluntarism was absolute. He believed that contingency extended all the way down to the "necessary truths" of logic and mathematics. "The mathematical truths, which you call eternal, were established by God and totally depend on him just like all the other creatures" ("Letter to Mersenne," quoted in Harrison (2002, 3)). Others, like Leibniz and Newton, held that God's goodness and wisdom were involved in these creative decisions at least in part and that God did not simply invent logic and mathematics by decree. There was more consensus, however, when it came to the laws of nature themselves. Most agreed with Boyle that the "laws of motion did not necessarily spring from the nature of matter, but depended on the will of the divine author of things" ([1690] 1772, 521) and were therefore contingent.

Notice that there are two ways of understanding contingency and necessity here, as Peter Harrison has pointed out (2002, 6). In contemporary modal logic, 'necessary' is contrasted with 'possible,' rather than 'contingent.' It is possible that the first letter in this paragraph is a capital T, since I could have begun the paragraph with the word 'The.' And since things could have been otherwise, it is not necessary that the first letter in this paragraph is a capital N (or T, C, etc.). In ancient and medieval philosophy, 'necessary' was contrasted with 'contingent.' A being is contingent when it depends on something else for its existence. Contingent beings are dependent; necessary beings are not. In Plato's system, the Forms were necessary. They were the foundation of reality and the source of being for everything else. A triangle drawn on a piece of paper is contingent in this sense. It depends on the paper and ink for its existence. The Form of Triangularity itself is a necessary being. In a theistic system, God is the only necessary being.[7] Everything else in creation depends on God for its existence. When early scientists influenced by voluntarism said that the laws of nature were not necessary, they meant it in both senses. Consider Newton's law of gravity, in modern notation,

$$F = G\frac{m_1 m_1}{r^2}$$

Roughly, it says that the force of gravity between two objects is a function of their two masses and the square of the distance between them. G is a constant. For Newton, this law is not necessary in both of the senses mentioned here. To say that this law is not necessary in the first sense (possibility) means that it might have taken a different form. G, for example, might have had a different value than it does in the actual world. To say that it is not necessary in the second sense (dependence) means that the existence of the law depends on the will of God. That there is a law of gravity is contingent on God's free choice, not his intellect or nature.

1.2.4　Experiments and Philosophy

The idea that nature might have been radically different posed an epistemic problem. If God had many, perhaps even infinitely many, combinations of laws from which to choose, how can we know which ones he actually did create? Reason alone cannot tell us, since it was God's will rather than reason that determined the laws. To understand the problem, contrast the voluntarist view with Plato's story of creation. In the *Timaeus*, a godlike craftsman known as the Demiurge wants to put the material world in order, to bring forth a cosmos from the chaos. The principles of order are found exclusively in the Forms, which are eternal and independent of the Demiurge. If the Demiurge is going to create anything, it must be modeled after the Forms. In a sense, the blueprints are already written and the Demiurge has no power to alter them. Hence, according to Plato, knowledge of the Forms is knowledge of the fundamental principles used to structure the cosmos and reason is the primary means for achieving this. The theistic God, on the other hand, is not subject to anything beyond himself. There are no eternal Forms. On the voluntarist (as opposed to Aristotelian/ Thomist) view, nothing channels God's decisions one way rather than another. Even with perfect knowledge of God's essence, one could still not infer what choices he made in creation.

This prompted another important change from medieval thought. As we noted earlier, experimentation was suspect on the Aristotelian view. Our innate ability to abstract universals from a set of particulars put the emphasis on reason, rather than hands-on trials. But if reason could not derive the laws of nature from first principles, then empirical investigation was the only option. In order to know what choices God actually made, natural philosophers would have to go see for themselves. As Nicolas Malebranche argued,

> It is certain that in this case one cannot discover the truth except by experience. For since we can neither grasp the designs of the creator nor understand all the relations which he has to his attributes, whether to conserve or not to conserve a constant absolute quantity of movement seems to depend on a purely arbitrary decision by God, about which we cannot become certain except by a species of revelation, such as is given by experience. (Milton 2003, 699)

Similar ideas were raised by mathematician Roger Cotes in the preface of the 2nd edition of Newton's *Principia*:

> Without all doubt this world, so diversified with that variety of forms and motions we find in it, could arise from nothing but the perfectly free will of God directing and presiding over all.
>
> From this fountain it is that those laws, which we call the laws of Nature, have flowed, in which there appear many traces indeed of the most wise contrivance, but not the least shadow of necessity. These therefore we must not seek from uncertain conjectures, but learn them from observations and experiments. ([1687] 1962, ii)

Those relying too much on preconceived views—especially Greek views—about perfection, creation, and God's choices in creation were subject to criticism. Kepler's thesis that the orbits of the planets conformed to the five Greek solids was rejected by Marin Mersenne on the basis that the divine will had many more possibilities open to it (Brooke 1991, 26). Kepler, Mersenne believed, ought not presume that God would use some particular principle. Only empirical investigation could discover the truth.

Before leaving the history of science, we should note that the influence of voluntarism on experimental science is more complex than might be suggested here. While Galileo was an experimentalist, his method was not motivated by voluntarism. In his view, the divine will was not arbitrary. There was a necessity in nature such that if we could discover all physical truths, we would see "that it would be impossible for them to take place in any other manner. For such is the property and condition of things which are natural and true" ([1632] 1953, 424). The difference between Galileo and his Aristotelian predecessors was that he believed this necessity was rooted in mathematics rather than in essences. Insofar as the truths of physics could be known, their implications could be derived with certainty. Finding these truths, however, required detailed observations, especially when showing that rival beliefs did not fit the data. The point here is that voluntarism was not a necessary condition for Galileo's empiricism.

More surprisingly, voluntarism did not always entail a move toward empiricism. Although Descartes was an extreme voluntarist, he is famously lumped in with the Continental Rationalists (with Spinoza and Leibniz) as opposed to the later British Empiricists (Locke, Berkeley, and Hume). In his view, God had stamped the fundamental truths of logic, mathematics, and at least some of the laws of nature in our minds (Harrison 2002, 4). Since

God would not deceive us—a well-known argument from his "Method of Doubt" (*Meditations* III.38)—we can trust the clear and distinct ideas that have been given to us. Careful reasoning was the Cartesian path to the first principles of natural philosophy, not experiments.

With these qualifications in place, there is still an undeniable connection between God's contingent choices in creation and the need for observation. Many, like Boyle and Newton, denied that science contains any necessary truths known *a priori*. Even in the case of Descartes, while his voluntarism did not lead to empiricism with respect to the *laws* of nature, things are different when it comes to the *mechanisms* in nature. God had many choices when it came to the configurations of material systems. Speaking about the size of particles in the universe and their behavior, Descartes says,

> Since there are countless different configurations which God might have instituted here, experience alone must teach us which configurations he actually selected in preference to the rest. We are thus free to make any assumption on these matters with the sole proviso that all the consequences of our assumption must agree with our experience. (*Principles of Philosophy* 3.46)

So while there is only one set of laws and these can be known *a priori* for Descartes, there are many different ways those laws can be implemented. Only observation can tell which God has chosen.

In short, voluntarism and nominalism did not logically entail the rise of empiricism. Nonetheless, theology played an important role in the creation of modern science. Theism was not simply tacked on to an otherwise naturalistic universe. What philosopher Daniel Garber concludes about Descartes could have been said about many others:

> In trying to link Descartes' physics closely to mathematics, one forgets the crucial connection between Descartes' physics and his metaphysics; it is a crucial feature of his physics that it is grounded in God, and without that grounding there could be no Cartesian physics. (1992, 293)

While experimental science might have arisen some other way, history shows that it emerged from within a theological framework. Once ancient Greek ideas about essences were rejected, there were religious reasons for believing that (i) nature is an artifact, a creation, governed more or less immediately by God; (ii) humans are able to learn the principles, laws, and mechanisms instituted by God; and (iii) observation is an essential part of attaining this knowledge.

1.2.5   The Galileo Affair

Not every chapter in this story is a harmonious one. After all, what about the persecution of Galileo? Surely, there was some tension between science and religion in the early modern period. Since this episode is referred to so often, let's consider it a bit more fully.

The first thing to note was the intellectual climate at the time. Exploration of the Americas had only recently begun. The Renaissance had undermined the hegemony of Aristotelian thought, and the Protestant Reformation had decreased the power of the papacy in favor of regional rulers. Political and theological tensions were unusually high. In the midst of this, Copernicus finally allowed the publication of his *On the Revolutions of the Heavenly Spheres* in 1543. Many astronomers at the time considered the Copernican model to merely be a computational device rather than a realistic picture of the cosmos. This was not Galileo's view. By 1611, he had constructed an improved telescope and began using it in an unconventional way: studying heavenly bodies. The prevailing view for centuries was that the celestial realm was perfect and unchanging, in contrast to the decay of the terrestrial sphere. The discovery of transient sunspots did not fit this picture, nor did the rocky craters observed on the moon—a very terrestrial-looking feature. Most damaging for geocentrism was the discovery that Jupiter has its own moons. Not all celestial bodies revolve around Earth.

Galileo's early work received high praise, especially among the Jesuits (Langeford 1998, 41–56), although he was opposed by Aristotelians who still held sway in the universities. Things began to change after the circulation of the *Letter to Castelli*, which was expanded in 1615 into the well-known treatise *A Letter to the Grand Duchess Christina*.[8] There, Galileo presented his own view of the relation between natural philosophy and biblical studies. This created a stir in that a Catholic scientist was now openly dabbling in matters of biblical interpretation, a realm outside of his area of expertise and into that of the clergy. In particular, there was a worry that Galileo was unintentionally lending support to a Protestant hermeneutic that gave too much weight to personal judgment. This was not a groundless accusation, as one can see in the *Letter to the Grand Duchess*:

> I do not feel obliged to believe that that same God who has endowed us with senses, reason, and intellect has intended to forgo their use and by some other means to give us knowledge which we can attain by them. He would not require us to deny sense and reason in physical matters which

are set before our eyes and minds by direct experience or necessary dem-
onstrations. ([1615] 1957)

Appealing to the autonomy of individual reason, especially in matters of
biblical interpretation, was a move associated with Luther rather than the
Pope. It didn't help matters that heliocentrism was somewhat better received
in Protestant circles than Catholic ones (Brooke 1991, 98). In light of this
perceived threat, Copernicanism was declared to be heretical by the
Inquisition in 1616. This put the Catholic Galileo on the wrong side of
things. Until there was proof to the contrary, he was directed not to promote
heliocentrism as fact but merely as a model that "saves the appearances"—a
view which is wrongly attributed to Copernicus himself.[9]

Galileo made a political miscalculation in 1632 with the publication of
the *Dialogue Concerning the Two Chief World Systems*. The dialogue had
three characters: an Aristotelian, a Copernican, and a layman. Rather
than a balanced discussion of the arguments and evidence, the book was
seen as Copernican propaganda and—rightly or wrongly—thought to be
making fun of Pope Urban VIII. This accusation was based on the fact
that his Aristotelian character in the dialogue, with the unflattering name
'Simplicio', presents an argument for geocentrism that the Pope himself
had previously used in discussions with Galileo. In 1633, Galileo was
found guilty by the Inquisition of teaching heliocentrism without proof.
He spent most of the remainder of his life confined to his villa outside of
Florence.

Unlike the flat earth myth, it is a fact that Galileo was condemned by the
Catholic Church. However, historians do not see the dispute as primarily a
matter of science versus religion as it is so often portrayed. Some, like John
Hedley Brooke, believe that the root of the conflict was between old science
and new science:

> Certainly the Catholic Church had a vested interest in Aristotelian philos-
> ophy, but much of the conflict ostensibly between science and religion turns
> out to have been between new science and the [old] sanctified science of the
> previous generation. (1991, 37)

In other words, while the Church had in hindsight placed its bet on the
wrong horse, the conflict had far more to do with change to an established
theory rather than religion. Had there not been opposition from university
Aristotelians trying to protect their cosmology, Galileo would have had no

one at whom to target his rhetoric. Things might have turned out differently had passions not run so high on both sides. Other scholars see the conflict as political:

> Historians have shown that the Galileo affair, remembered by some as a clash between science and religion, was primarily a dispute about the enduring political question of who was authorized to produce and disseminate knowledge. (Dixon 2008, 31)

This certainly characterized the dispute by the time *Two Chief World Systems* was published. The Church could not afford to see one of its most prominent scientists making public arguments that could be wielded by the Protestant opposition.

Careful editing can allow one to spin the historical record in several different ways. Many prefer the narrative that the scientific revolution was the liberation of science and rational thought from religion. More recently, some have argued for the other extreme that Christianity alone was able to give birth to modern science.[10] Historians of science now reject all such oversimplifications. Slogans cannot capture the influence of religion on the rise of modern science.

The goal here was to provide enough background to discuss the overall relationship between science and religion. Before we get to that, though, let's first consider the structure of science itself in a bit more detail.

1.3   The Structure of Science

There is no simple picture that captures the nature of science. Scientific knowledge is too diverse to fit under a single principle, even the so-called "scientific method." With that in mind, let's consider a model that is useful as a first approximation.

1.3.1   Three Layers

Broadly speaking, science has three layers (Figure 1.1). The first is the level of observations and data. Like a pyramid, this is the broadest layer in the sense that there are more data available than theories or models to place them in. This level includes both careful observations of natural phenomena as well as experimentation and controlled studies.

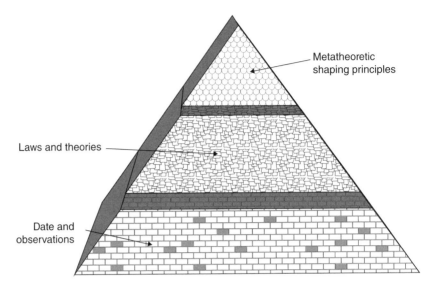

Metatheoretic shaping principles

Laws and theories

Date and observations

Figure 1.1   Three layers of science.

The second layer is more abstract, containing theories and laws. Toward the bottom of this layer, relatively close to the data itself, are statistical correlations and phenomenological models. As statisticians repeatedly tell us, establishing a correlation is not the same as discovering a cause, but it is an important first step. We know that high cholesterol is correlated with heart disease, but drugs that lower cholesterol haven't dramatically affected the rate of cardiac deaths in America. The complete causal story appears to be more complex than "high cholesterol causes heart disease." Phenomenological models are used to replicate patterns found within data. They are built in a "bottom-up" fashion in the sense that there are no first principles or laws of nature from which to derive them. For example, with enough data, one can create a computer model that simulates the behavior of city traffic, even though there are no general equations governing traffic flow. Higher up in the second layer are more abstract laws and mature theories including Einstein's field equations and relativity, Schrödinger's equation and quantum mechanics, and the nonlinear differential equations used in statistical mechanics, continuum mechanics, and chaos theory.

The least familiar layer is at the top, the level of metatheoretic shaping principles (MSPs).[11] This is the region where the philosophy of science and science proper blend into one another. There is no sharp line between the two. MSPs help determine what good theories, laws, and models look like as well as how one should proceed in their discovery and development. Philosophers of science have considered the role of these principles for more than a generation, and there are several ways to approach the subject. I will divide such principles into two categories. Note that while these are commonly assumed in modern science, only a few are without controversy in the philosophy of science. The first are metaphysical and include:

- *The primacy of laws.* The universe is governed by a set of regularities, the laws of nature.

Although this idea is no longer tied to theism, much of the history of modern physics involves the discovery, refinement, and replacement of what these laws are believed to be.

- *Uniformity of nature.* This is uniformity across space and time. The laws of nature are thought to be the same now as they always have been, or at least since the earliest stages of the universe. The laws of nature are the same here as they are everywhere else in the universe.

If this were not so, few sound inferences could be made in astrophysics or geology.

- *Causation.* Every physical event has a sufficient physical cause, except possibly at the level of quantum mechanics.
- *Realism.* Mature theories in science embody discovered truths about reality. Theories are not merely social constructions.

Realism is usually supported by the "no-miracles" argument: it would be a miracle if science could be as successful as it has been and not be more or less true. While this principle is debated among philosophers, it is taken for granted in most areas of the natural sciences.

The second category of MSPs is epistemic and includes a wide variety of methodological norms.

- Reliance on *repeatable, intersubjective observations.* As we have seen, this is one of the principles that sets apart modern from medieval science.
- Standards of *inductive logic and mathematical rigor.* This includes the proper use of statistical methods and blind studies.

- *Explanatory virtues.* These are desiderata for good explanations including empirical adequacy, simplicity, testability, internal and external coherence, fruitfulness for future research, wide scope, and elegance.

Most of the explanatory virtues are familiar ideas. Coherence has to do with how well parts of a theory fit together with each other (internal) and with the rest of established science (external). Scope is the breadth of phenomena the explanation covers. Newton's laws, for example, had extraordinarily large scope, covering microscopic bodies ("corpuscles"), everyday objects, as well as the planets. Testability means that the explanation is sufficiently concrete that observations could either confirm or disconfirm it.[12] Elegance is usually a mathematical aesthetic, one that was surprisingly useful in twentieth-century physics. A lack of fruitful research, in contrast, is often interpreted as a dead end. In his criticism of ray optics in the nineteenth century, for example, Whewell complained that "there is here no unexpected success, no happy coincidence, no convergence of principles from remote quarters; … this is not the character of truth" (Whewell 1837, 2:428). While none of these are necessary conditions for an acceptable explanation, any theory that cannot claim several of these virtues is in trouble. Continuing with epistemic MSPs:

- *Conservatism.* As new discoveries are made, scientific theories should change as little as needed in order to accommodate them.

This is closely related to W.V.O. Quine's doctrine of "minimal mutilation": new observations may force a change in one's beliefs, but one should make the smallest change possible in order to accommodate the new information (1980, 42–44).[13]

- *Tenacity.* Good theories are difficult to displace. They earn the right of continued acceptance even in the face of some anomalies.

Tenacity is in part a king-of-the-hill doctrine in epistemology: one should keep one's current set of beliefs unless something better comes along to displace them. The mere fact that there are other possible views, even equally good ones, ought not be enough to change one's justified beliefs. Tenacity is in stark contrast to the stereotype that in "science, we can abandon at the drop of a hat beliefs that we have held dear for centuries, once we have new information" (Ecklund 2010, 108). Most changes do not come so easily.

When organic chemist Herbert C. Brown challenged the accepted model of carbonium ions several decades ago, he recalls that his thesis was treated as "a heresy, triggering what appeared to be a 'holy war' to prove me wrong" (Brooke 1991, 18). Notice that in this case tenacity inhibited a legitimate scientific advance. While that's always a risk, allowing good theories to earn the right not to be overthrown at the first sign of trouble also reduces faddishness and unwarranted change.

- *Methodological naturalism.* Science can only appeal to natural laws and physical entities as explanations of observable phenomena.

This MSP will be of special interest in later chapters, especially in matters of divine action and the intelligent design (ID) debate.

1.3.2   Change and Suspension

Each of the MSPs mentioned is found in modern science, but many others have been changed or set aside. One is that nature works only by contact forces, like the gears in a clock. On this view, scientific explanations should primarily describe the mechanism responsible for the phenomena. Newton famously violated this principle with his theory of universal gravitation. Although the equation for the force of gravity was widely praised, Newton refused to posit a mechanism to which this law applied. Without something material that could be manipulated by contact, Newtonian gravity was criticized for relying on action at a distance, something akin to telekinesis. Eventually, the demand for contact forces was given up.

Some of the most interesting conflicts in the history science arise over which shaping principles can legitimately be challenged. One example is the Bohr–Einstein dispute over quantum mechanics. According to Niels Bohr's view of quantum mechanics—commonly referred to as the Copenhagen interpretation—nature is fundamentally random. Some physical events have no sufficient cause.[14] Thus, the Heisenberg uncertainty principle is not merely a limitation on our knowledge; it is a metaphysical truth—the way things are. Einstein saw this as a radical and unwarranted change. To Einstein, Bohr was advocating the overthrow of law-governed causal regularities as they had been understood from the beginning of the scientific revolution. Einstein argued instead that Heisenberg uncertainty merely puts a limit on predictions. It tells us nothing about what facts of the matter exist regarding the position and momentum of a particle. Bohr won this dispute.[15] Most interpretations of quantum mechanics include an irreducible element of chance, limiting the MSP of causal determinism.

A less famous eighteenth-century debate over shaping principles involved mathematician/physicists Leonhard Euler and Jean le Rond d'Alembert.[16] Like Descartes before him, d'Alembert believed that mathematics was a highly specialized tool that could only describe simple systems. Much of what we observe in the physical world is beyond its resources. Even something as common as a plucked string was thought to be too complex to be described by differential equations. Euler was less strict. He argued that the rules governing the use of differential equations should be relaxed on occasion, even if it meant ignoring a fundamental metaphysical doctrine, Leibniz's law of continuity. The law of continuity says that "nature makes no leaps"; the change from one system state to the next is always continuous. Leibniz himself thought this principle was a cornerstone of mechanics, arguing that "continuity [is] a necessary prerequisite or a distinctive character of the true laws of the communication of motion. [Can] we doubt that all phenomena are subject to it … ?" (1702, Letter to Varignon quoted by Crockett (1999, 120)). Nonetheless, Euler ignored Leibniz's law in his analysis and mathematical physics has successfully followed his lead ever since.

Another major dispute over shaping principles soon followed. Early geologists held a view now called *catastrophism*: most geological structures are the result of large-scale events such as floods and earthquakes. Both the temperature and surface of the Earth were thought to have changed dramatically over time in the wake of massive, sporadic natural disasters. Among the supporters of catastrophism was the founder of comparative anatomy and vertebrate paleontology, Georges Cuvier. A rival view known as *uniformitarianism* was developed by James Hutton[17] and later entrenched in Charles Lyell's *Principles of Geology* (1830). Uniformitarians taught that geological data should be explained only in terms of continuous, ongoing forces and mechanisms such as the slow rise of mountains, underground cooling of magma, rain, erosion, and sedimentation, all extending back through time:

> In examining things present, we have data from which to reason with regard to what has been; and, from what has actually been, we have data for concluding with regard to that which is to happen hereafter. Therefore, upon the supposition that the operations of nature are equable and steady, we find, in natural appearances, means for concluding a certain portion of

time to have necessarily elapsed, in the production of those events of which we see the effects. (Hutton [1788] 2007, 16)

The older catastrophic view was attacked on two fronts. The first had to do with new data, namely, the discovery of fossils. The similarity between fossilized and living creatures supported continuity between the past and the present. Change over time was incremental, with gradual sedimentation explaining the correlation between types of fossils within specific geological strata. The second came from other shaping principles that had taken hold by that time. Catastrophic explanations were out of sync with the mechanistic picture of nature and its continuously acting laws. Unlike the slow, ongoing processes of Hutton and Lyell, ancient catastrophes were unobservable and hence thought to be beyond the reach of empirical study. As such, the older view was deemed unscientific. In fact, Lyell went so far as to argue that ad hoc appeals to floods, earthquakes, and the like were on a par with the view that demons are responsible for moral failures (Anderson 2007, 452–453).[18]

There are three important points to note in all this. First, like everything else in science, MSPs can be modified, suspended, and even rejected outright when there are sufficient reasons to do so. Second, shaping principles both influence the development of theories and are themselves influenced by changes in other layers of the pyramid model of science. In fact, influences flow from each of the three layers to every other layer. New observations and anomalies force changes in theories, but currently held theories also affect the way one evaluates new data. Whether an observation is significant, for example, depends on what theories one already accepts.[19] Further up the pyramid, both observations and new theories can put pressure on shaping principles. The Einstein–Bohr debate discussed earlier is an example of this sort of pressure. Third, appealing to the "scientific method" for guidance in all this is futile. Any candidate for the scientific method rests on some notion of what it is to be good science, a notion which is itself grounded in MSPs. Insofar as the scientific method depends on shaping principles, it cannot be used to determine what those principles should be.

1.3.3   Religion and Shaping Principles

One drawback of the pyramid model is the suggestion that science is self-contained, as if the three layers affect each other, but there are no links to any other realm of thought. Even our brief historical overview shows that this is not the case. Religion and philosophy (as well as sociology and politics) have shaped the development of science, and we can now say where this influence normally occurs: at the level of MSPs. Religious beliefs seldom have direct influence on theories and even less on observations. Shaping principles are a different story. The rationality of nature, the law-governedness of nature, and early conservation principles were all motivated by beliefs about the choices God had made in creation. This isn't to say religion has always been helpful. Catastrophism was tied to a young earth creationist interpretation of the Old Testament. The main "catastrophe" in this theory was the Flood of Noah. Uniformitarians had to contend with both the theoretical and biblical arguments posed by catastrophists before modern geology could move forward. And although the reality of the Galileo affair is rather different than commonly portrayed, the Church's endorsement of Aristotelian metaphysics certainly hindered the acceptance of heliocentrism.[20]

All this points to an unavoidable and unappreciated truth: science is messy. Change in any of the three layers can arise unexpectedly and the implications must be worked out over time.

Having covered some of the history and philosophy of science, we can now look more closely at the overall relationship between science and religion. First, some qualifiers. When the term 'religion' is used here, it will refer mostly to religious beliefs. There is, of course, much more to religion than propositional knowledge, but for our purposes, we can ignore this. Likewise, there is more to science than what scientists publish in textbooks and journal articles. Science is also about experimentation and methods. Insofar as these are important to our discussion, I take them to be captured by MSPs. In other words, if double-blind experiments *should* be used by medical researchers when possible, this practice can be described by MSPs as having to do with what good science looks like in medicine. In general, the virtues and vices of scientific practice will be subsumed here under the epistemological MSPs.

1.4   The Relation between Science and Religion

Dating back to Ian Barbour's work in the 1960s, the issue is usually framed as finding "the correct model" for the relation between science and religion. Barbour gives what have become the four standard answers. While many scholars now think this typology is too lean, it's still a useful starting place.

1.4.1   Conflict/Warfare

While it's well and good that religion had a positive influence on the early history of science, proponents of the conflict model (Figure 1.2) stress that the harmony has not lasted. Conventional wisdom says that the two now exist in a state of conflict, often with organized religion hindering scientific progress. One finds this theme everywhere from embryonic stem cell research to Dan Brown novels.[21] The Scopes "Monkey Trial" (1925), at least in its *Inherit the Wind* portrayal, is a paradigm case of religious ignorance resisting scientific thought. In fact, the controversy over Darwinian evolution has been the backbone of the conflict model since the 1860 debate between Bishop Samuel Wilberforce and Thomas Huxley. An account written ten years later recalls the scene:

> Then the Bishop rose, and in a light scoffing tone … assured us there was nothing in the idea of evolution; rock-pigeons were what rock-pigeons had always been. Then, turning to his antagonist with a smiling insolence, he begged to know, was it through his grandfather or his grandmother that he claimed his descent from a monkey? On this Mr Huxley slowly and deliberately arose. … He [answered that he] was not ashamed to have a monkey for his ancestor; but he would be ashamed to be connected with a man who used great gifts to obscure the truth. … I, for one, jumped out of my seat; and

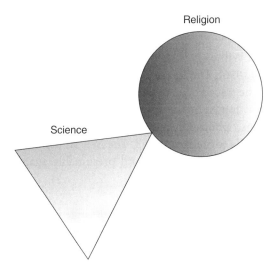

Figure 1.2   Conflict.

when in the evening we met at Dr Daubeney's, every one was eager to con-
gratulate the hero of the day. (Sidgwick 1898, 433–434)

Huxley's biographer later described the debate as nothing less than an "open
clash between Science and the Church" (Livingstone 2009, 155).[22] Popular
accounts of science and religion often repeat this narrative of rationality
over dogmatism.

The ideas behind the conflict model took root during the European
Enlightenment (late 1600s to early 1800s). Among its better known advocates were the French writer Voltaire, the great Scottish philosopher David
Hume, and the American revolutionary Thomas Paine. Consider this
passage from *The Age of Reason*:

> The persons who first preached the christian system of faith, and in some
> measure combined with it the morality preached by Jesus Christ, might per-
> suade themselves that it was better than the heathen mythology that then
> prevailed. … But though such a belief might, by such means, be rendered
> almost general among the laity, it is next to impossible to account for the con-
> tinual persecution carried on by the church, for several hundred years,
> against the sciences, and against the professors of science, if the church had
> not some record or tradition that it was originally no other than a pious
> fraud, or did not foresee that it could not be maintained against the evidence
> that the structure of the universe afforded. (Paine 1794, chap. XVI)

The conflict model was explicitly promoted in such books as John W.
Draper's *History of the Conflict between Religion and Science* (1874) and
Andrew Dickson White's *A History of the Warfare of Science with Theology
in Christendom* (1896). The flat earth myth mentioned at the beginning of
this chapter can be traced back to these two authors.[23]

Conflict is now touted by the so-called "New Atheists" such as biologist
Richard Dawkins, philosopher Daniel Dennett, and physicist Victor
Stenger. The subtitle to Stenger's book, *The New Atheism*, is *Taking a Stand
for Science and Reason*—taking a stand presumably against religion and
irrationality:

> The position of the New Atheists is that faith is the force behind both the
> malevolent deeds of extremist religious groups and the irrational acts of
> many political leaders. To act on the basis of faith can often be to act in
> conflict with reason. We New Atheists claim that to do so is immoral, and
> dangerous to society. (Stenger 2010, 12)

The power of science is

> based on self- and mutual criticism and a humble acceptance of uncertainty
> in our conclusions … [while] religion is on the contrary blatantly arrogant
> in its unselfcritical commitment to unfounded certainties and dogmas.
> (Stenger 2010, 14)

Science versus religion in this context is portrayed as part of the larger
conflict between reason and faith: the scientific and educated on one side
and the religious and dogmatic on the other. The idea that the two are
founded on conflicting epistemologies is a common theme in this literature. Chemist Peter Atkins expands on Stenger's critique:

> A scientist's explanation is in terms of a purposeless, knowable, and … fully
> reduced simplicity. Religion, on the other hand, seeks to explain in terms of
> a purposeful, unknowable, and incomprehensible irreducible complexity.
> Science and religion cannot be reconciled. (Haarsma 2010, 110)

This inability to reconcile the two leads to conflict. Most proponents of the
conflict model happily agree with Stephen Hawking: "Science will win"
(Heussner 2010, 1).

Not everyone who holds this model is antagonistic toward religion, however. A number of conservative Christians have come to accept conflict as a
consequence of young earth creationism. If the Bible teaches that the Earth
is less than 20 000 years old and yet geologists say 5–6 billion, there is no
way to reconcile the two. Creationists believe that modern science conflicts
with what they take to be a more reliable source of information regarding
the origin of the universe. More precisely, they believe the problem is with
modern, textbook science, not future or ideal science. In other words, the
tension is not a matter of principle based on the nature of science and theology, but rather a temporary conflict having to do with the scientific consensus as it stands today. Many creationists believe that their views will be
vindicated in the future. Others are less optimistic about such a reconciliation. In any case, young earth creationists accept the conflict model vis-à-
vis current theories of geology, cosmology, and evolutionary biology.

When it comes to scholars who specialize in the study of religion and
science, on the other hand, few, if any, accept the conflict model. One reason
is that, as we have seen, it doesn't fit the historical record nearly as well as
nineteenth-century authors such as Draper and White claimed. Even

Galileo—often portrayed as the poster boy for this view—does not offer much support once the historical details are considered. Another reason is the long list of religious scientists who have detected no tension between their theological beliefs and their vocations. Among them, Michael Faraday's religious views are well known,[24] and James Clerk Maxwell saw no inconsistency in including an argument for the existence of God in his *Encyclopædia Britannica* entry on atoms (1870).[25] According to a recent study, even today, almost half of the elite scientists in America are "religious in a traditional sense," and another 20% have "spiritual sensibilities that often derive from and are borne out in the work they do as scientists" (Ecklund 2010, 6, 130). While none of this entails that such beliefs are true, it does shift the burden of proof to advocates of the conflict model. Perhaps these scientists are self-deceived or have failed to recognize a deep incongruity between the two discourses. If so, conflict theorists are welcome to make that case. For now, the opinion of scholars lies elsewhere: "The greatest myth in the history of science and religion holds that they have been in a state of constant conflict" (Numbers 2009, 1).

Why then is this view so popular? One is that the media has a strong preference for "us versus them" narratives: Democrats versus Republicans, pro-choice versus pro-life, liberals versus conservatives, and reason versus faith. These tropes make it easy for the audience to recognize a particular fight and to identify which side to root for. Another is that there *is* often a real conflict going on in the background for which science versus religion is a proxy. The actual tension is between supernatural theism and metaphysical naturalism.[26] Naturalism says that everything that exists is natural stuff, and natural scientists are the ones who tell us what that is. Immaterial entities such as essences, souls, angels, and God are ruled out. Naturalism and theism are obviously incompatible, since naturalism entails atheism. But science is not synonymous with naturalism nor is religion only theism. While science influences our metaphysics, metaphysics cannot be reduced to science, or at least it would require some argument in order to believe that it does. Many advocates of the conflict model simply assume that (i) the only things we are entitled to believe to exist are those things that scientists study and that (ii) science is the only real source of knowledge. Unfortunately, these claims lie somewhere between "highly controversial" and "hopelessly naive." Most of the knowledge obtained and transmitted in universities today falls outside the realm of science. And scientists themselves hold metaphysical views that are not entailed by any theory or observation, as our discussion of MSPs showed. In any case, the point here is that while there is a legitimate conflict between theism and naturalism, that debate ought not be conflated with religion versus science.

1.4.2 Independent Realms

Many academics agree that there are vast differences between science and religion but believe that conflict is the wrong conclusion. To see why, notice first that many factual beliefs have little or nothing in common. Lawn care and astrophysics are completely different concerns. Advances in baking have nothing to do with discoveries in paleontology. For conflict to be possible, there must first be some sort of overlapping subject matter. On the independence model (Figure 1.3), science and religion do not intersect. Religious beliefs do not impinge on physical theory in any way, and so friction between their domains is impossible. Hence, neither side need feel threatened by the other. The two realms are epistemically autonomous.

The most famous version of this model was put forward by the American paleontologist Stephen J. Gould. He called it *nonoverlapping magisteria* (NOMA):

> [The magisterium] of science covers the empirical realm: what is the universe made of (fact) and why does it work this way (theory). The magisterium of religion extends over questions of ultimate meaning and moral value. These two magisteria do not overlap.... (Gould 1999, 6)

The two magisteria thus cover completely different areas within the human experience. Science does not address questions about meaning, purpose, and value. Religion does not address matters of fact about the natural world.

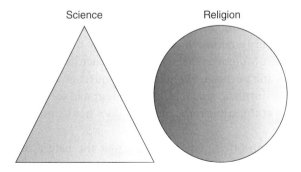

Science   Religion

Figure 1.3   Independence.

Many scientists like this model because it effectively keeps religious matters at arm's length. Scientists are sometimes nervous about religion and worry about any attempt to influence science in inappropriate ways, especially when it comes to creationism. But if religion properly understood has no say regarding empirical matters, then theological issues can be safely ignored. Anyone raising religious concerns misunderstands the independence of science, and scientists need not waste time addressing their complaints. This is one reason why critics of ID stress that it is a religious point of view rather than a scientific one. If correct, then scientists can happily ignore ID arguments.

Agnostic scientists are not the only advocates of the independence model. Theists have held some version of it since the German theologian Friedrich Schleiermacher in the early 1800s. Schleiermacher believed that religion needed to be protected from natural science. One way to do this was to deny that religion is a matter of knowledge. Schleiermacher held that religion is instead based on feeling, most importantly a feeling of complete dependence upon God. It has nothing to do with the claims of empirical science. Scientific thought is about wholly different matters; it is irrelevant to religion. This autonomy meant that religion would be safe from science-based attacks like those forthcoming from Freudians, Marxists, and Darwinists. No matter what science might discover, religious faith would carry on. With no overlap between the two, atheists cannot argue that naturalistic science disproves theism any more than the success of quantum mechanics means one should be an Ohio State fan. There is no logical connection between them.

While the independence model remains popular in some circles, most experts take it to be oversimplified. Both science and religion are far more complex than Gould and others allow. First, while Gould characterizes religion as the domain of value and purpose, science is not value-free. Controversies involving plagiarism, proper credit for discoveries, climate change, genetically engineered crops, and many more are at least in part matters of ethics. Moreover, the practice of science itself presupposes MSPs about its goals and purpose. There is also the question of whether scientific knowledge has intrinsic value or is only desirable for the technology it yields. Whatever the answer, the question does not seem to be a matter of religion.

Second, religion often makes claims about the nature of the physical world and observable events. If miracles ever occur, as most theists believe, then these are events in the natural world.[27] The New Testament story of

Jesus raising Lazarus from the dead is purportedly a historical event. The point here is not to argue for its historicity, but rather that the matter is not merely about ethics, value, and purpose. Gould himself did not consider miracles to be a counterexample to NOMA since they improperly mix the two magisteria and were therefore ruled out of bounds. But this simply begs the question. Independence is easy to prove if one can conveniently rule out counterexamples that do not fit the model. In any case, religion also makes ontological claims, as Richard Dawkins points out (1997). The existence of souls has to do with the nature of human beings, a belief that Dawkins argues is at odds with common ancestry. And whether God exists is a metaphysical question, not a moral one. That religion makes such claims shows that it is about more than matters of value and purpose.

So far, our options have been limited: either conflict and culture war on one hand or enforced segregation and harmony on the other. Both models engage in cherry-picking, selecting those examples that support their view and ignoring the rest. While the next one is not as simple, it captures more of the nuances seen in the history of science.

1.4.3   Dialogue

On the dialogue model (Figure 1.4), science and religion can and sometimes should influence each other in a variety of ways. Scientific knowledge can be of use in theology and, as history shows, religion can have a positive influence on science. Some will find it ironic that Galileo himself argued for a view along these lines. Many natural philosophers at the time believed that God authored two books: the book of scripture and

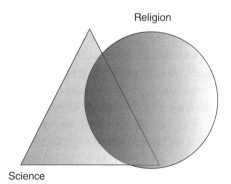

Figure 1.4   Dialogue.

the book of nature. Religion is the means by which one studies the first book; science is the means by which one studies the second. While God ensures that there is no inconsistency between the two books themselves, there is no such guarantee for our fallible interpretations of those books. Moreover, Galileo argued that divinely inspired texts use language that, although appropriate for the original audience, was often imprecise and phenomenological, hence the need for both books. In the *Letter to the Grand Duchess*, Galileo says,

> [The] holy Bible and the phenomena of nature proceed alike from the divine Word, the former as the dictate of the Holy Ghost and the latter as the observant executrix of God's commands. It is necessary for the Bible, in order to be accommodated to the understanding of every man, to speak many things which appear to differ from the absolute truth so far as the bare meaning of the words is concerned. But Nature, on the other hand, is inexorable and immutable; she never transgresses the laws imposed upon her.... For that reason it appears that nothing physical which sense-experience sets before our eyes, or which necessary demonstrations prove to us, ought to be called in question (much less condemned) upon the testimony of biblical passages which may have some different meaning beneath their words. ([1615] 1957)

Copernicus himself, said Galileo, "did not ignore the Bible, but he knew very well that if his doctrine were proved, then it could not contradict the Scriptures when they were rightly understood" ([1615] 1957). So while the Bible talks about the sun rising, setting, and miraculously standing still, the book of nature shows that this language merely reflects how things look from our point of view. The second book provides insight into what the first book teaches and what it does not.

More recent versions of the dialogue model agree that science sometimes has a role in correcting religious assumptions. For centuries, it was believed that the celestial and terrestrial realms were completely different; Newton showed that one set of laws governs both. Abrahamic religions taught that the universe was created less than 20 000 years ago, while others said that it was infinitely old. Geology and cosmology show that both are wrong. But the dialogue is not all in one direction. It was from theism that science adopted shaping principles regarding the rationality and law-governedness of nature. Religion might also provide explanations for scientific brute facts, such as why gravity is precisely as strong as it is (more on that in Chapter 2). Shaping principles for the aims and value of science can be influenced by religious concerns. And finally, as philosopher Alan Padgett (2010) notes, monasteries, hospitals, and universities have been vital for the development and transmission of scientific knowledge since the Middle Ages. This was possible only because their religious founders believed this knowledge was worth pursuing.

While this is widely viewed as the best of the four standard models in the science and religion literature, let's briefly consider the fourth before we get too far into the analysis.

1.4.4   Integration

Some academics have proposed that we work to integrate all knowledge into one coherent whole. Not only might there be a unified theory of everything in physics, but all of scientific and religious understanding should be fit together—ecumenicalism on the highest possible scale. This view is not new. Early advocates of integration such as Charles William Elliot, president of Harvard in 1877, were caught up in the effervescence of progress: "In every field of study, in history, philology, philosophy, and theology, as well as in natural history and physics, it is now the scientific spirit, the scientific method, which prevails" (Sloan 1994, 3). Others, like clergyman and natural philosopher Joseph Priestley, wanted a more rational Christianity that could stand up to Enlightenment critiques. Priestley believed that God would use science to purge theology of superstition. More recently, integration has been a consistent theme within process theology.[28]

The integration model has been the least popular of the four. Those hostile to uninterested in religion prefer to keep the two realms separate. Others are concerned that science and theology are never equal partners once the integration schemes start getting fleshed out. This is because the typical way of achieving synthesis is to naturalize religion to one degree or another. Doctrines that cannot be easily integrated with contemporary science are simply discarded. Priestley himself wanted to discard the doctrine of the Trinity, which could be written off as the influence of Platonism on the early church (Brooke 1991, 25). Biologist and Anglican priest Arthur Peacocke believed that the virgin birth of Jesus must be rejected on the grounds that genetics requires chromosomes from both male and female parents. Many Christians reacted with puzzlement. Could not an omnipotent God find a way around this problem? In short, if integration means that all supernatural aspects of religion are done away with, most consider the price too high.

The hope for integration has also suffered from the demise of reductionism in science. Reductionism says that entities at one level of reality are nothing but composites or states of entities at a more fundamental level. On this view, minds and mental states are nothing but brains and brain states; brains and brain states are nothing but neurons, nerve fibers, and their states, and so on all the way down to molecules, atoms, and fundamental particles, whatever they happen to be. In terms of the relation between the sciences, reductionism says that as science progresses, we should expect higher-level theories to be explained in terms of more fundamental theories. Macroevents and laws should become explicable in terms of those at a lower level, much the way thermodynamics has (supposedly) been reduced to statistical mechanics.

The plausibility of full-scale reduction across the sciences is now extremely doubtful (more on this in Chapter 6). If anything, science has become more specialized and fragmented in recent decades. Even some of the paradigm cases of reduction are problematic in the details. Thermodynamics, for example, has in fact never been fully reduced to atomic physics. While reduction might continue to be successful in specific, isolated cases, the grand reductionist hope is essentially dead and with it, so it would seem, is the integration model. If the branches of physics are not themselves fully integrated with each other, let alone the whole of science, then the idea that scientific and religious knowledge might one day fit together in one package is implausible.

Finally, as Mikael Stenmark argues, integration is merely an extreme version of dialogue in which the interaction eventually produces one system of thought (Stenmark 2004, 251–252, 260–269). Hence, integration does not warrant a category of its own.

1.4.5   Assessment

While each of these models has flaws, the four-part typology itself has recently come under criticism, as if these are the only four ways science and religion can be related. Even the Venn diagrams in Figures 1.2, 1.3, 1.4, and 1.5 impose a misleading sense of mathematical precision. Matters are not that clear cut for three reasons. First, both disciplines are broad and diverse. The variety of religions can sometimes be overwhelming, so that part should be no surprise, but science also has tremendous breadth. Consider what little paleontology has in common with computational chemistry. Even within physics itself, there are vast differences between high-energy

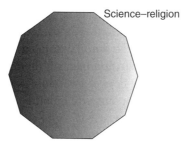

Figure 1.5   Integration.

physics and continuum mechanics in terms of content and between the the-
oretician and the experimentalist in terms of methodology (and some, such
as astronomers, who don't fit neatly into either of those categories).

Second, neither science nor religion has clear boundaries. Is religion
everything that is of "ultimate concern," as theologian Paul Tillich contended?
If so, the circle of religion is quite large including many people who would
not consider themselves religious. Perhaps religion should be limited to the
worship of a divine being. Then again, classical Buddhism is usually listed
among the world's religions, so perhaps not. There seems to be no precise
fact of the matter as to what counts as religion. Some examples are clearly in
(e.g., Judaism), some are clearly out (e.g., plumbing), and others are border-
line cases that could go either way (e.g., Confucianism). Science also has a
fuzzier boundary than normally assumed. This is one reason philosophers
have been so skeptical about the existence of a strict "scientific method"—one
used by all the sciences and only the sciences (Laudan 1983). There are no
criteria for what counts as science that seem to include all of the good examples (general relativity, biochemistry, archeology, etc.) while ruling out the
bad ones (alchemy, astrology, etc.). One of the issues regarding ID is whether
it is science or religion. (One can usually tell advocates from critics based on
which side they put ID. We will see why this is a more difficult example than
commonly assumed in Chapter 5.) MSPs also exist along the fuzzy border
between science and philosophy. While I put shaping principles on the science side of the ledger, they are often instead considered matters of metaphysics and epistemology that are necessary conditions for science. This is a
judgment call insofar as the border between the two is not well defined.

Third, as we noted earlier, science is not self-contained. Religion has
played a role, but so have sociology and politics. Science is not merely a
social construction (contrary to what exegetes of philosopher Thomas

Kuhn's work would have us believe). It is, however, a human enterprise shaped in part by interpersonal relationships and intellectual fads. Much of physicist Lee Smolin's *The Trouble with Physics* is a diagnosis of how string theory has unfortunately come to dominate his field. Too much of the story, in Smolin's view, is due to the heavy-handed influence of key individuals and physics departments.[29] The point here is that sociological influences are going to be part of any institution; science is no exception.

In short, it's going to be difficult to have a simple model that correctly states how science and religion are related if (i) we can't say precisely what science and religion are and (ii) the two are shaped by so many outside influences.

Each of the four models we've considered treats as precise matters that instead lack precision, like saying where the atmosphere ends and space begins. In some cases, science and religion conflict. Young earth creationism is an example. In most ways, the two are independent. Whether matter is atomic, continuous, or quantum mechanical has nothing to do with the essential properties of God. There is no one-size-fits-all model. The debate over *the* correct model depends very much on what one wants to emphasize and what one wants to downplay or ignore. Historians of science now urge a different approach:

> [S]erious scholarship in the history of science has revealed so extraordinarily rich and complex a relationship between science and religion in the past that general theses are difficult to sustain. The real lesson turns out to be the complexity. (Brooke 1991, 5)

Instead of discrete models, Stenmark argues for different axes along which science and religion might converge or diverge. These include (i) goals, (ii) the means of achieving those goals (methodology), and (iii) the products of applying those means (content) (Stenmark 2004, 260). One might argue for some degree of either contact or independence along each of these coordinates depending on how they are interpreted. For example, one form of independence points to differing goals: science is exclusively about explanations and predictions about the physical world; religion is mostly about explanations and descriptions of the supernatural realm and matters of value and character. Someone arguing for contact could instead claim that the goal of both is truth, most generally truth about the ways things are. Hence, science and religion might overlap as they pursue the same goal.

Again, there's a lot of cherry-picking here. Advocates of the conflict model emphasize different methods, science based on repeatable observation and testing of hypotheses and religion on personal experience, the authority of scriptures, and bodies of experts. Proponents of the dialogue model instead point to the fact that both use *inference to the best explanation* (aka abduction). Roughly, this rule of inference says that the best explanation for a given event is most likely true, where "best" is often determined by way of the explanatory virtues mentioned in Section 1.3.1. Physicians use inference to the best explanation when weaving a patient's symptoms into the most coherent explanation to form a diagnosis. Prosecutors use it to connect evidence to suspects. Scientists use this rule in theory selection and in the positing of unobservable entities. One can't see electrons, but positing their existence allows us to explain electromagnetic and chemical phenomena. One can't go back in time and observe Aztecs, but positing their existence allows us to explain a range of artifacts.[30] Likewise, many theists argue that although God is unobservable, positing the existence of such a being allows us to explain religious experiences, miracle claims, ethical truths, etc. Those who prefer the dialogue model can therefore say there is methodological overlap between science and religion at a sufficiently abstract level since both make use of inference to the best explanation.

Stenmark's multiple axes are certainly helpful compared to the standard models. Four categories cannot capture the nuance of the positions one finds in the literature. He also rightly points out that the relation between science and religion changes over time. There is no doubt more independence today than had been in the early modern period. We should therefore resist imposing a single relation between the two that supposedly works for all time. The only weakness in Stenmark's analysis is that it is almost wholly descriptive, allowing us to get a better handle on the positions in this debate without giving any guidance as to what the right answer is. The open question is how we *should* think about science and religion today, if in fact there are any helpful templates to be found.

1.4.6   Proposal

By now, we should agree that there are no bumper sticker slogans for the relation we seek. What then can we say? The reader might have noticed that I often call science and religion 'disciplines.' That is no accident. It was once commonly assumed that while there is one reality "out there," there is a division of labor in the university when it comes to studying that reality.[31] Some disciplines are closely related, like molecular biology and biochemistry, while others are rather distant, such as classics and computer engineering. Each discipline typically makes progress on its own set of issues happily ignoring goings-on elsewhere. On the other hand, some topics require expertise from many sources. If we want to understand poverty, we will need input from experts in economics, sociology, history, political science, and psychology—and that's the short list.

Recall the earlier qualifier that 'religion' here refers to religious beliefs. We're ignoring religious practices in all this. With that restriction, I believe the best way to think of religion and science is as independent disciplines generally pursuing their own questions with their own methods. Religion in this sense includes systematic theology, biblical studies, and a host of subdisciplines. Science is composed of the social and natural sciences and all of their specializations. Atheists and metaphysical naturalists will object that theology does not in fact study any aspect of reality and should not be considered an academic discipline; it is an empty discourse in which terms like 'God' fail to denote anything that exists. Although I believe that's wrong, we do not need to resolve those questions here. I am going to assume, along with most philosophers of religion, that theism is a rational position. Perhaps one day, orthodox theology, string theory, and adaptationism will all be considered dead ends. Each has its detractors, and anyone who thinks that a given research program will fail is free to ignore it. But so long as enough academics are betting the other way, each should be treated as a live option.

Academic disciplines typically address their own issues with distinct methods appropriate to their subjects. Most—but certainly not all—of what happens in one disciple is irrelevant to the others, even closely related ones. For science and religion, then, we shouldn't expect "dialogue" on every issue or even many issues. This also means that conflicts will be rare. The metaphysical debate over naturalism will continue, but that is a conflict between philosophical positions, not disciplines. Integration is even less plausible. While interdisciplinary work between organic chemists and molecular biologists is of some use, no one thinks that integrating biology and chemistry is a good idea. The two have different methods and research agendas.

One ramification of this proposal is that the search for *the* model for the relation between science and religion is misguided. Do we need a model for the relation between biology and physics or between psychology and economics? Other than the question of whether one field can be reduced to another (see Sections 6.2 and 6.3), the answer is no. We look for insight wherever we can find it. Interdisciplinary crossover can be helpful, often in unforeseen ways, but there is no need for an overarching model of how disciplines fit together. Science and religion are no different in this regard.

My proposal is not a call for more workshops and conferences bringing scientists, philosophers, and theologians together. As we've noted, most researchers go about their own business most of the time. My view gives scientists permission not to engage religion. Some argue this is precisely the wrong message to send; scientists need more encouragement to understand religion, not arguments allowing for even more separation (Ecklund 2010, 112). While I sympathize with this, the fact is that while crossover can be encouraged, many academics do not find interdisciplinary work helpful. High-energy physicists often ignore solid-state physics and dismiss continuum mechanics as merely phenomenological. Matters only get worse when it comes to the interaction between physics and chemists or chemists and biologists or any of the sciences and philosophy. But for those who *are* interested in this sort of interdisciplinary work, it should be supported and might again prove to be helpful as it has in the past. This is not so much a radical new perspective as a return to the thinking of past generations of scientists, including Einstein:

> I fully agree with you about the significance and educational value of methodology as well as history and philosophy of science. So many people today— and even professional scientists—seem to me like somebody who has seen thousands of trees but has never seen a forest. A knowledge of the historic and philosophical background gives that kind of independence from prejudices of his generation from which most scientists are suffering. This independence created by philosophical insight is—in my opinion—the mark of distinction between a mere artisan or specialist and a real seeker after truth. (Letter to Robert Thornton, December 7, 1944)

Like all interdisciplinary work, there is the potential for both conflict and fruitful interactions. The rise of heliocentrism was unusual in that observations refuted the then current view among theologians—and nearly everyone else, for that matter.[32] What the Galileo affair shows is that discoveries in science can spur reevaluation of religious points of view, even scriptural interpretation. Some on the religion side will find this disconcerting, but if science helps show that the reasons for believing a given view were poor in the first place, as was the case with geocentrism, then this is a valuable exchange.

Most interactions between the two disciplines will be more subtle. First, there are metaphorical extensions: finding a useful idea in one field that is indirectly applicable in another. Chaos theory has many such extensions from its home in nonlinear dynamics to law, psychiatry, nursing, and literature.[33] This isn't to say that the systems in those disciplines are governed by differential equations or that strange attractors can be detected in their evolution. Rather, there are ideas in chaos, most notably sensitive dependence on initial conditions (i.e., the butterfly effect), that many find useful outside of the natural sciences. Similarly, Descartes' principle of inertia was in part a metaphorical extension from God's immutability and his creation/continual preservation of matter (Brooke 1991, 75). Leibniz's max/min principles in optics were directly influenced by the principle of perfection in creation (Milton 2003, 698). (Some have also argued that Maxwell's view of the electromagnetic field was a metaphorical extension of his reflections on the Trinity (Torrance 2001, 14–15); however, the textual evidence for this claim is weak.)

There are also metaphorical extensions flowing back the other way, from science to religion. One is the notion of *complementarity*. Bohr used the term to describe the apparent contradictions that arise from subatomic entities behaving as particles in some experiments and waves in others. Although it seems that light must either be a wave or a group of particles, Bohr argued that both descriptions are needed to fully understand the phenomena. There is no higher synthesis possible. One must simply come to grips with the dual nature of light and matter. Bohr came to apply this idea far beyond quantum mechanics, even suggesting that it might be used to understand the tensions between science and religion (MacKinnon 1996, 266). Although the two might seem contradictory at times, both are needed to understand the whole of reality. Finally, paleontologist Pierre Teilhard de Chardin used a metaphorical extension from evolution to theology. His Omega Point doctrine held that not only were biological systems on Earth evolving, but so was the entire universe, bringing forth its potential for greater degrees of complexity and rationality. The entire process is being drawn to its culmination by the Omega, the ultimate Aristotelian final cause, something that Teilhard de Chardin believed "is suggested, but not proved, by scientific analysis" (McGrath 1999, 224).

The second major way the two disciplines influence each other is by serving as a source of shaping principles. As we've seen, that the universe is governed by laws of nature was historically linked to theistic creation. So was the view that those laws apply universally, as Newton argued in an early draft of the *Opticks*:

If there be an universal life and all space be the sensorium of a thinking being who by immediate presence perceives all things in it [then] the laws of motion arising from life or will may be of universal extent. (Ratzsch 2010, 65)

The MSP of simplicity also had theistic roots, as we saw previously in Descartes.[34] Newton likewise held that God "had ensured that nature did nothing in vain," and Faraday believed that the book of God's works "would be as simple to comprehend as the books of his words" (Brooke 1991, 28). And while Einstein's famous quip "God does not play dice" was not a reference to the theistic God, it did reflect a theological/metaphysical influence on his thinking about causation and determinism.

Finally, religion can on rare occasions tip one's opinion in favor of one scientific theory over a rival. This happened with Faraday's rejection of billiard ball atoms in favor of centers of force—something along the line of Boscovich's point masses. He argued that God's creative ability made one atomic theory more plausible than another.[35] Science can have a reciprocal effect when it comes to competing doctrines in theology. Chaos, relativity, and quantum mechanics have all influenced doctrines involving God's relationship to time, matters of determinism and free will, and the theology of miracles and divine action. Each of these will be discussed more fully in later chapters.

The big worry scientists have in encouraging this sort of interdisciplinary work is that religion will encroach in unwelcome ways, with creationism and ID theory leading the way. I believe such concerns should be taken seriously. How can we tell when tension is due to legitimate interdisciplinary research as opposed to the infiltration of theological doctrines attempting to change science? (Note that infiltration works the other way as well, when metaphysical naturalism is dressed in the garb of science in order to "demythologize" religion.) These questions will be taken up in Chapter 5 when we deal with ID and naturalism in more detail. For now, I advise patience in such matters. One can seldom say in advance when interdisciplinary crossover will pay off. Who would have thought that chaos theory, grounded in classical mechanics, would be applicable in cardiology and ecology, that continuum mechanics would be important to marine biology, or that theism might be relevant to atomism? In any case, interdisciplinary studies are never forced on researchers; one engages other fields for personal interest or to see if there are untapped resources in other departments. For scientists who aren't interested, let them be. But for those who are, such crossover should not be forbidden. If a given research program is fruitless, that will become evident in time. Let a thousand flowers bloom, as the saying goes.

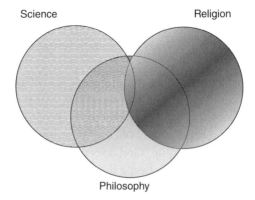

Figure 1.6   Three disciplines.

I've called the interdisciplinary view "my proposal," but in many ways, it just describes what's going on in the philosophy of religion and philosophy of science these days.[36] Philosophers of religion are interested in any apparent conflict between theology and science. Philosophers of science try to tease out the ramifications of scientific theories—what they tell us about reality and how we come to know it. As a discipline, philosophy overlaps both science and religion. By its nature, it tends to be interdisciplinary (Figure 1.6).

So regardless of whether specialists within science and religion, respectively, are interested in exploring points of contact, philosophers of science and religion certainly are. Arguments about science and the existence of God, the nature of space and time, free will and foreknowledge, the laws of nature and divine action, and design and evolution have been part of philosophy for centuries.

With the stage now set, let's consider these arguments in order.

Notes

1   The one notable exception in the west was the sixth-century monk, Cosmas Indicopleustes.

2   The full English title of Newton's *Principia* was *Mathematical Principles of Natural Philosophy*. The subject matter of natural philosophy included the study of physical reality, but went far beyond that, even to "the attributes of God and His relationship to the physical world" (Brooke 1991, 7).

3   For the sake of simplicity, I'm ignoring the distinction between *potentia absoluta* and *potentia ordinata*. Strictly speaking, most medievals would agree that God could *de potentia absoluta* do anything that does not violate the principle of noncontradiction. Once having decided to create a cosmos with a range of essences, however, God limits himself to act accordingly within the structure of that order, *de potentia ordinata*. The latter is constrained by God's wisdom, the former is not. So while the break from Aristotelian necessity was accelerated by voluntarism, it began in the intellectualist tradition which included Aquinas.

4   Roberto Torretti mentions that it was used in medical contexts to describe living a healthy lifestyle and by the Greek Church Fathers in talking about such things as cycle of birth and death (Torretti 1999, 405–406).

5   Also see Oakley (1961). As Milton also shows, the idea of a law of nature had been widely used as a theory of ethics for several centuries prior to its use in natural philosophy (2003, 681). See, for example, Aquinas's *Summa Theologica* II.1 q90–94.

6   Although Ted Davis argues that the influence of voluntarism comes in degrees. From Galileo to Descartes to Boyle, there is increasing reliance on voluntarist ideas (Davis 1999).

7   Technically, God is the only *uncreated* necessary being (necessary *a se*). Some metaphysicians recognize a type of created yet necessary being, necessary *ab alio*. I ignore that complication here.

8   Although widely circulated, the full version was not officially published until 1636 (Langeford 1998, 58).

9   An antirealist interpretation of the heliocentric model was promoted in an unsigned foreword to *On the Revolutions of the Heavenly Spheres* by Andreas Osiander, a Lutheran theologian. Since Copernicus died at the time of publication, he could not correct the notion that this was his own view.

10 For a critique of both of these extremes, see Numbers (2009), especially Chapters 8–10.

11 The name 'shaping principle' was coined by philosopher Del Ratzsch. See chapter 7 of Ratzsch (2001).

12 Philosophers of science no longer believe that scientific explanations must be *falsifiable*. If one is willing to put up with enough *ad hocness*, any explanation can be preserved in light of disconfirming evidence. Testability is a more modest idea. It means that a good scientific explanation "must be put *in empirical harm's way*" (Ratzsch 2001, 98). Experiment and observation must be able to add or detract from its plausibility. As physicist Julian Barbour puts it, "The fact is that, in a choice between two different theoretical schemes, there must always be a preference (in the absence of compelling experimental evidence) for the one that is more restrictive and, hence, makes stronger predictions" (Barbour *et al.* 2002, 24).

13   What Quine actually said in "Two Dogmas of Empiricism" was that minimal
      mutilation is "our natural tendency"—a descriptive claim. In the hands of phi-
      losophers of science like Larry Sklar, minimal mutilation became normative
      (Sklar 1975). Belief change should be minimal; dramatic changes ought to be
      considered only when necessary. This is now a widely accepted view among
      epistemologists.

14   This will be discussed more fully in Chapter 4.

15   Don Howard argues that the core of this debate was over quantum entangle-
      ment, not causation and determinism per se. "[E]ntanglement, not indeterminacy,
      was the chief source of Einstein's misgivings about quantum mechanics. …
      Indeterminacy was but a symptom; entanglement was the underlying disease"
      (http://www.science20.com/don_howard/revisiting_einsteinbohr_dialogue).
      Whichever point of contention drove this debate, MSPs played a major role.

16   This episode has been well documented by mathematician/historian Clifford
      Truesdell (1984, 80–83) and philosopher Mark Wilson (2000, 298–301).

17   Hutton's rejection of catastrophism was clear in his first edition of his *Theory
      of the Earth*: "Philosophers observing an apparent disorder and confusion in
      the solid parts of this globe, have been led to conclude, that there formerly
      existed a more regular and uniform state, in the constitution of this earth; that
      there had happened some destructive change; and that the original structure
      of the earth had been broken and disturbed by some violent operation, whether
      natural, or from a supernatural cause. Now, all these appearances, from which
      conclusions of this kind have been formed, find the most perfect explanation
      in the theory which we have been endeavouring to establish; for they are the
      facts from whence we have reasoned, in discovering the nature and constitution
      of this earth: therefore, there is no occasion for having recourse to any unnatural
      supposition of evil, to any destructive accident in nature, or to the agency of any
      preternatural cause, in explaining that which actually exists" ([1788] 2007, 63).

18   Ironically, Walter Alvarez and David M. Raup, the leading supporters of the
      asteroid impact hypothesis as a major cause of mass extinction, attribute the stiff
      resistance they faced in the 1970s to uniformitarianism (Raup 1999, 35–36).

19   The technical name for this is *theory-ladenness*, and there is far more to this
      issue than I have alluded to here. Since the 1960s, many have argued that the
      influence of currently held beliefs on new data undermines realism and the
      objectivity of science. For an introduction, see chapter 6 of Kosso (1992).

20   Protestant hermeneutics didn't help matters. Martin Luther argued that the
      Bible disproves heliocentrism since "Joshua bid the sun to stand still and not
      the earth" (*Table Talk* Number 2387 a–b).

21   "Since the beginning of history," Langdon explained, "a deep rift has existed
      between science and religion. Outspoken scientists like Copernicus—" "Were
      murdered," Kohler interjected. "Murdered by the church for revealing scientific
      truths. Religion has always persecuted science." (Brown 2000, 31)

22 Most historians take this and the Sidgwick account to be a bit of revisionist history regarding the 1860 British Association for the Advancement of Science meeting. See, for example, Dixon (2008, 73–76) and Livingstone (2009).

23 Washington Irving's *A History of the Life and Voyages of Christopher Columbus* (1828) was also responsible for the flat earth myth, insofar as Columbus is portrayed as bravely risking the lives of his crew to disprove it.

24 This is evident even in his earliest biography, J.H. Gladstone's (1873). Faraday himself argued that "God has been pleased to work in his material creation by laws, and these laws are made evident to us by the *constancy* of the characters of matter and the *constancy* of the effects which it produces" (Levere 1968, 105).

25 "The period of vibration of a luminous particle is … a quantity which in itself is capable of assuming any one of a series of values, which, if not mathematically continuous, is such that consecutive observed values differ from each other by less than the ten thousandth part of either. There is, therefore, nothing in the nature of time itself to prevent the period of vibration of a molecule from assuming any one of many thousand different observable values." Maxwell's explanation for this was the same as Sir John Herschel. This uniformity is the mark of engineering. Atoms "must therefore have been made."

26 Physicist Deborah Haarsma makes a similar point but identifies the sides as religion versus reductive materialism (2010, 111). 'Materialism' is an old term for the metaphysical position that all that exists is material stuff, such as atoms. In the twentieth century, it became clear that the physicist's ontology included more than atoms. Philosophers therefore dropped 'materialism' in favor of 'physicalism': everything that exists is physical stuff, whatever the physicists tell us that is. Today, the still broader term 'naturalism' is often used: everything that exists is natural stuff.

27 The nature of miracles will be discussed more fully in Chapter 4.

28 Also known as *panentheism*, process theology takes the physical universe to be contained in God without being identical to God, as it is in pantheism. One analogy is that God's relationship to the universe, according to process thought, is like a mind's relationship to its body: thoroughly interactive, but not identical.

29 Smolin's experience is an interesting commentary on the sociology of science: "[There is] a sense of entitlement and a lack of regard for those who work on alternative approaches to the problems string theory claims to solve. … [The] major string theory conferences never invite scientists working on rival approaches to give papers. This of course serves only to bolster the assertions of string theorists that string theory is the only approach yielding successful results on quantum gravity. The disregard of alternative approaches sometimes borders on disdain. At a recent string theory conference, an editor from Cambridge University Press confided to me that a string theorist had told him he would never consider publishing with the press because it had put out a book on loop quantum gravity. This kind of thing is not as rare as it should be.

... If you raise detailed questions about one of string theory's claims with an expert, you risk being regarded, with faint puzzlement, as someone who has inexplicably chosen a path that precludes membership in the club. Of course, this isn't true of the more open-minded string theorists—but there is a peculiar tightening of the face muscles that I've seen too often to ignore, and it usually happens when a young string theorist suddenly realizes that he or she is talking to someone who does not share all the assumptions of the clan" (Smolin 2006, 270–271).

30 Philosophers who take a realist approach to science tend to be the strongest advocates of inference to the best explanation. Antirealists think this is misguided. See, for example, van Fraassen's criticism in his work (Van Fraassen 1989, 131–150).

31 The rise of postmodernism has put that in question. Some, like philosopher Richard Rorty, have argued that the notion of a single, mind-independent reality about which we try to gain knowledge should be rejected. They believe that there is no God's-eye point of view from which to discuss reality as it is in itself, what Immanuel Kant called the "noumenal realm." This will be taken up in more detail in Chapter 7. For now, I simply assume a generally realist perspective.

32 As Robert Bishop rightly points out (private correspondence), Tycho Brahe complicates matters somewhat. Brahe proposed a geocentric model in which the other planets orbited the sun. Galileo's data was consistent with this model as well as the Copernican one.

33 Philosopher Stephen Kellert (1995) coined the term 'metaphorical extension' in his work on chaos theory in the 1990s.

34 "The reason for this rule is the same as the reason for the first rule, namely the immutability and simplicity of the operation by which God preserves matter in motion" (Descartes, *Principles of Philosophy* 2.39).

35 "What real reason, then, is there for supposing that there is any such nucleus in a particle of matter? ... Is the lingering notion which remains in the minds of some ... that God could not just as easily by his word speak power into existence around centers, as he could first create nuclei & then clothe them with power?" (Faraday, "Matter," 1844 in Levere (1968, 107)).

36 My proposal is a close cousin to philosopher Alan Padgett's "mutuality model" (2003).

References

Anderson, Owen. 2007. "Charles Lyell, Uniformitarianism, and Interpretive Principles." *Zygon* 42 (2): 449–462.
Barbour, Julian, Brendan Z. Foster, and Niall Ó. Murchadha. 2002. "Relativity Without Relativity." *Classical and Quantum Gravity* 19 (12): 3217–3248.

Boyle, Robert. [1690] 1772. "Christian Virtuoso." In *The Works of the Honourable Robert Boyle*, edited by Thomas Birch, 5:508–540. London: J & F Rivington.

Brooke, John Hedley. 1991. *Science and Religion: Some Historical Perspectives*. Cambridge: Cambridge University Press.

Brown, Dan. 2000. *Angels & Demons*. New York: Atria Books.

Cotes, Roger, and Isaac Newton. [1687] 1962. "Preface." In *Principia Mathematica*, edited by Florian Cajori, translated by Andrew Motte. Vol. 1, xxxii. Berkeley: University of California Press.

Crockett, Timothy. 1999. "Continuity in Leibniz's Mature Metaphysics." *Philosophical Studies* 94 (1–2): 119–138.

Davis, Edward B. 1999. "Christianity and Early Modern Science: The Foster Thesis Reconsidered." In *Evangelicals and Science in Historical Perspective*, edited by D.N. Livingstone, D.G. Hart, and M.A. Noll, 75–95. New York: Oxford University Press.

Dawkins, Richard. 1997. "Science Discredits Religion." *Quarterly Review of Biology* 72: 397–399.

Deason, Gary B. 1986. "Reformation Theology and the Mechanistic Conception of Nature." In *God and Nature: Historical Essays on the Encounter Between Christianity and Science*, edited by David C. Lindberg and Ronald L. Numbers, 167–191. Berkeley: University of California Press.

Dixon, Thomas. 2008. *Science and Religion: A Very Short Introduction*. New York: Oxford University Press.

Ecklund, Elaine Howard. 2010. *Science vs. Religion: What Scientists Really Think*. New York: Oxford University Press.

Galilei, Galileo. [1632] 1953. *Dialogue Concerning the Two Chief World Systems, Ptolemaic & Copernican*. Translated by Stillman Drake. Berkeley: University of California Press.

Galilei, Galileo. [1615] 1957. "Letter to the Grand Duchess Christina." In *Discoveries and Opinions of Galileo*, edited by Stillman Drake, 175–216. New York: Anchor Books.

Garber, Daniel. 1992. *Descartes' Metaphysical Physics*. Chicago: University of Chicago Press.

Gladstone, John Hall. 1873. *Michael Faraday*. London: Macmillan and Company.

Gould, Stephen J. 1999. *Rocks of Ages: Science and Religion in the Fullness of Life*. New York: Ballantine.

Haarsma, Deborah B. 2010. "Science and Religion in Harmony." In *Science and Religion in Dialogue*, edited by Melville Y. Stewart, 1:107–119. 2 vols. New York: Wiley-Blackwell.

Harrison, Peter. 2002. "Voluntarism and Early Modern Science." *History of Science* 40: 63–89.

Heussner, Ki Mae. 2010. "Stephen Hawking on Religion: 'Science Will Win.'" *Abcnews.com*. http://abcnews.go.com/WN/Technology/stephen-hawking-religion-science-win/story?id=10830164. Accessed March 19, 2012.

Hutton, James. [1788] 2007. *Theory of the Earth*. Sioux Falls: Nuvision Publications.

Kellert, Stephen H. 1995. "When Is the Economy Not Like the Weather? The Problem of Extending Chaos Theory to the Social Sciences." In *Chaos and Society*, edited by A. Albert, 35–47. Amsterdam: IOS Press.

Kosso, Peter. 1992. *Reading the Book of Nature: An Introduction to the Philosophy of Science*. Cambridge: Cambridge University Press.

Langeford, Jerome J. 1998. *Galileo, Science and the Church*. South Bend: St. Augustine's Press.

Laudan, Larry. 1983. "The Demise of the Demarcation Problem." In *Physics, Philosophy and Psychoanalysis: Essays in Honor of Adolf Grünbaum*, edited by L. Laudan and R.S. Cohen, 76:111–127. Doordrecht: D. Reidel.

Levere, Trevor H. 1968. "Faraday, Matter, and Natural Theology: Reflections on an Unpublished Manuscript." *British Journal for the History of Science* 4 (2): 95–107.

Livingstone, David N. 2009. "That Huxley Defeated Wilberforce in Their Debate over Evolution and Religion." In *Galileo Goes to Jail and Other Myths about Science and Religion*, edited by Ronald L. Numbers, 152–169. Cambridge: Harvard University Press.

MacKinnon, Edward. 1996. "Complementarity." In *Religion and Science: History, Method, Dialogue*, edited by Mark W. Richardson and Wesley J. Wildman, 256–270. New York: Routledge.

McGrath, Alister E. 1999. *Science and Religion: An Introduction*. Oxford: Blackwell.

McMullin, Ernan. 1967. "Empiricism and the Scientific Revolution." In *Art, Science, and History in the Renaissance*, edited by Charles S. Singleton, 331–369. Baltimore: Johns Hopkins Press.

Milton, J.R. 2003. "Laws of Nature." In *The Cambridge History of Seventeenth Century Philosophy*, edited by Michael Ayers and Daniel Garber, 1:680–700. Cambridge: Cambridge University Press.

Newton, Isaac. [1687] 1966. *Principia Mathematica*. Edited by Florian Cajori. Translated by Andrew Motte. Vol. 1. 2 vols. Berkeley: University of California Press.

Numbers, Ronald L. 2009. "Introduction." In *Galileo Goes to Jail: And Other Myths about Science and Religion*, edited by Ronald L. Numbers. Cambridge: Harvard University Press.

Oakley, Francis. 1961. "Christian Theology and the Newtonian Science: The Rise of the Concept of the Laws of Nature." *Church History* 30 (4): 433–457.

Padgett, Alan. 2003. *Science and the Study of God: A Mutuality Model for Theology and Science*. Grand Rapids: Eerdmans.

Padgett, Alan. 2010. "Science and Religion in Western History." In *Science and Religion in Dialogue*, edited by Melville Y. Stewart, 2:849–861. Malden: Wiley-Blackwell.

Quine, W.V.O., editor. 1980. "Two Dogmas of Empiricism." In *From a Logical Point of View*, 20–46. Cambridge: Harvard University Press.

Ratzsch, Del. 2001. *Nature, Design, and Science*. Albany: SUNY Press.

Ratzsch, Del. 2010. "The Religious Roots of Science." In *Science and Religion in Dialogue*, edited by Melville Y. Stewart, 1:54–68. New York: Wiley-Blackwell.

Raup, David M. 1999. *The Nemesis Affair*. New York: W.W. Norton.

Sidgwick, Isabella. 1898. "A Grandmother's Tales." *Macmillan's Magazine* 78 (468): 425–435.

Sklar, Lawrence. 1975. "Methodological Conservatism." *Philosophical Review* 84 (3): 374–400.

Sloan, Douglas. 1994. *Faith and Knowledge: Mainline Protestantism and American Higher Education*. Louisville: Westminster John Knox Press.

Smolin, Lee. 2006. *The Trouble with Physics*. Boston: Houghton Mifflin.

Stenger, Victor. 2010. "What's New About the New Atheism." *Philosophy Now: A Magazine of Ideas*, May.

Stenmark, Mikael. 2004. *How to Relate Science and Religion*. Grand Rapids: Eerdmans.

Torrance, Thomas F. 2001. *Theological and Natural Science*. Eugene: Wipf and Stock Publishers.

Torretti, Roberto. 1999. *The Philosophy of Physics*. Cambridge: Cambridge University Press.

Truesdell, C. 1984. *An Idiot's Fugitive Essays on Science: Methods, Criticism, Training, Circumstances*. New York: Springer-Verlag.

Van Fraassen, Bas C. 1989. *Laws and Symmetry*. Oxford: Clarendon Press.

Whewell, William. 1837. *History of the Inductive Sciences from the Earliest to the Present Time*. Vol. 2. 3 vols. London: John W. Parker and Sons.

Wilson, Mark. 2000. "The Unreasonable Uncooperativeness of Mathematics in the Natural Sciences." *The Monist* 83 (2): 296–314.

2

Fine-Tuning and Cosmology

2.1   What Is Fine-Tuning?

Tropical fish are beautiful to watch. If you think that fish are easy pets though, you should do some research before buying a tank. Many tropical fish are extremely sensitive to temperature, salinity, and the level of nitrogen, oxygen, and phosphates. These and a host of other environmental variables have to be just right to keep little Nemo from going belly up.

One of the most intriguing discoveries of 20th-century physics is that the universe is a bit like an aquarium. For life to be possible, two dozen or so cosmological variables have to be just right. Change any one by even a slight amount and living creatures could not exist here or anywhere else in the universe. That's not what scientists expected. They always knew that if the laws of nature were different, then living creatures would be changed as well. If the value of Newton's gravitational constant[1] had been slightly greater, for example, giraffes would probably not exist. There would be too much weight to support on spindly legs, and their hearts could not pump blood all the way to their brains. In general, vertebrates would have to be shorter and squattier with thicker bones in order to support the extra weight. What scientists didn't expect is that the precise value of the gravitational constant determines whether the universe can support any life at all. The surprising precision of these constants, such as the G in Newton's law of gravity, is known as *fine-tuning*. The universe shouldn't care whether life exists or not. Why then do so many of its fundamental parameters seem to be set to the precise values needed for our existence?

The Physics of Theism: God, Physics, and the Philosophy of Science,
First Edition. Jeffrey Koperski.
© 2015 John Wiley & Sons, Ltd. Published 2015 by John Wiley & Sons, Ltd.

Most physicists and philosophers believe that fine-tuning needs an explanation. Theism of course provides a ready answer: the universe looks fine-tuned for life because it has been fine-tuned for life. Our cosmic environment bears the earmarks of design. In this chapter, we will consider some examples of fine-tuning (Section 2.2), the best naturalistic explanations (Section 2.3), and whether the need for explanation is itself based on faulty premises (Section 2.4).

2.2 Examples

In the 30 years since fine-tuning was recognized, many papers and books have been written with detailed examples, so I won't be spending a lot of time on them here. The goal of this section is to convey a basic sense of what needs to be explained and then get on with the arguments. We will consider two types of fine-tuning, those dealing with the initial conditions of the universe and those based on fixed parameters.[2]

2.2.1 Initial Conditions

Descartes believed that universe would eventually end up more or less the way it is now regardless of how it started.[3] Like a boulder rolling down a steep incline, exactly where and how it began rolling doesn't matter. Eventually, it will land at the bottom. He was wrong. Physicists now believe that a slight change in the initial conditions immediately after the Big Bang would have had highly undesirable effects. Let's consider an example.

One of the most researched types of fine-tuning starts with the humble second law of thermodynamics. This law entails that all closed systems tend to evolve away from orderly configurations toward ones with high entropy. Fallen tree limbs never just form into neat stacks by themselves. Batteries do not spontaneously recharge. Entropy is itself a measure of disorder. A system in equilibrium has maximum entropy; there isn't any more usable energy left. We also know that the average entropy for all of observable space is low (most of the order is concentrated in stars and galaxies) and that the universe is billions of years old. Given the second law, the present state of the universe would have had to evolve from one with even lower entropy. That's fine, except for the fact that the early universe had far more high-entropy states available. In order to produce a universe like the one we see 14 billion years after the Big Bang, the earliest state of the universe must have had unimaginably low entropy—a very special state among those that were physically possible.

How special? Say we create a mathematical state space where each point in that space represents one possible state of the universe.[4] Each dimension of this space is one degree of freedom—one way in which the universe could vary (Figure 2.1).

Divide up (or "coarse grain") this space into a bunch of small boxes (Figure 2.2).

Now, put all of the points representing states with sufficiently low entropy to produce our observable universe in one, full box (Figure 2.3).

Figure 2.1   Universal state space (limited to 3 dimensions).

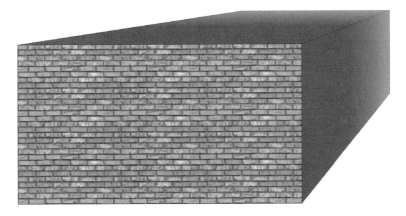

Figure 2.2   Coarse-grained space.

Figure 2.3 Low-entropy states.

If all of the boxes are the same size, how many total boxes would there be in that space? Mathematician Roger Penrose has famously calculated that this special box of low-entropy states would be one of $10^{10^{123}}$ such boxes (1989, 339–344). This is the largest physically significant number most people have ever encountered. If this box were chosen at random, then the odds would be far better for you to find one particular particle from all of the protons, electrons, and neutrons in the observable universe. Somehow nature chose the right box from all the initial states possible. All of the other boxes have too much entropy to support life.

Physicists and philosophers of science consider this to be one of the most pressing cases of fine-tuning in need of explanation.[5] Others, like philosopher Craig Callender, argue that initial and boundary conditions do not require explanations. We'll take up that sort of argument in Section 2.3.3.2.

2.2.2 Fixed Parameters

According to the Standard Model of particle physics, there are 20 or so free parameters whose values must be determined by experiment. There is no reason these constants have the magnitudes that they do. Their actual values are brute facts: a truth for which there is no further explanation. Brute facts are just the way things are. If so, then these parameters might just have well taken some other value. "[As] far as we can tell, the universe might have been created so that exactly the same laws are satisfied, except that the values of these parameters are tuned to different numbers" (Smolin 1999, 37).

The key mechanism at work here is known as *symmetry breaking*. Consider the rotational symmetry of a ping-pong ball. No matter how one rotates the ball sitting on a table, it looks the same as it did before. A perfectly cylindrical pencil balancing on its tip also has rotational symmetry; it looks the same from all sides. This symmetry is broken once the pencil falls. Now, the angles look rather different depending on one's vantage point. In the early universe, entropy was very low and energy levels were extraordinarily high. In these extreme conditions, which can be approximated in large particle accelerators, there was a symmetry between electromagnetism and the weak nuclear force. Instead of two forces, there was only one, dubbed the "electroweak" force, with an infinite range. Once the universe cooled to around 10^{15}°K, this symmetry was broken. Two distinct forces emerged with different properties. In particular, the coupling constants α and α_w were fixed with their current values, which determined the relative strength of these forces. Most physicists believe this same story repeats itself at even earlier stages such that the electroweak force was once merged with the nuclear strong force. Many are betting that shortly after the Big Bang, there was only one force and that through repeated symmetry breaking, we eventually end up with the four fundamental forces we have now: electromagnetism (α), the weak (α_w) and strong (α_s) nuclear forces, and gravity (α_g).

For our purposes, the most important part of this story is that the values of the four coupling constants and the sizes of the elementary particles associated with them could have been different. Just as the pencil has no natural proclivity to fall in one direction rather than another, these values could have been vastly different. From nature's perspective, there was no reason to prefer one set of values over another. This is where the fine-tuning comes into play.

2.2.2.1   The Nuclear Weak Force

If this force had been 30 times weaker, hydrogen would have been far easier to convert into helium after the Big Bang. This would have resulted in a prevalence of helium stars in the universe rather than hydrogen stars (Davies 1982, 63–65). Helium stars tend to be less stable and burn for shorter time, making them far less suitable as energy sources for life on nearby planets. Perhaps, more importantly, there would be much less hydrogen available to form water. If the weak coupling parameter α_w had been slightly stronger, neutrinos would react more readily with matter at the core of prenova stars. This would prevent them from delivering their energy to the outer layers. It is this very energy that drives a supernova.

Without it, large stars would not die in the explosive way they do and hence not disperse the heavy elements essential for life that were discussed in the previous section. Either way, a change in this force would produce a universe significantly less suitable for life.

2.2.2.2   The Nuclear Strong Force and Carbon Production

Let's consider how some of these essential-for-life elements were produced (Davies 2006, 134–135). First, there was abundance of hydrogen in the early universe. Helium comes from hydrogen via fusion in stars. Getting from helium to anything higher on the periodic table is more difficult. Adding a proton to a helium nucleus produces lithium-5, which is unstable. (Stable lithium requires more neutrons.) Two helium nuclei together produce beryllium-8:

$$_2^4\text{He} + {}_2^4\text{He} \rightarrow {}_4^8\text{Be}$$

which is also unstable with a half-life of 10^{-17} seconds. To find a stable nucleus, we need to get all the way to carbon. That would require three helium nuclei to collide at precisely the same time—a highly unlikely event. If this bottleneck weren't overcome, the universe would only have two elements in any significant amount. And yet here we are, in a universe with an abundance of carbon and more.

In the early 1950s, astronomer Fred Hoyle predicted that there must be an energy resonance that facilitates the production of carbon. Wave resonance makes energy transfer easier, which is what happens when an opera singer breaks a glass once reaching a certain pitch. According to quantum mechanics, particles also have wave properties. Hoyle correctly predicted that carbon has a resonant state such that it would be relatively easy to form a $_6^{12}\text{C}$ nucleus if only the right energy could be achieved. Experiments showed that the "right" energy was 7.656 MeV. With this resonance and some help from the thermal energy of the star, beryllium-8 can quickly react with helium to produce carbon and some excess energy (Barrow 2002, 153):

$$_4^8\text{Be} + {}_2^4\text{He} \rightarrow {}_6^{12}\text{C} + e^+ + e^-$$

Happily, we cannot tell a similar story for oxygen, which has a resonant state at 7.1187 MeV. When carbon combines with helium, the resulting energy level is 7.1616 MeV—too high for resonance. That's good, since otherwise most of the carbon produced in stars would become oxygen. As it is, things are just right for both.

All of this hinges on the precise value of α_s. Oberhummer *et al.* reported in the journal *Science* that a 0.4% increase or decrease in the nuclear strong force would wreck the balance we have now (Oberhummer *et al.* 2000). Varying this constant either way "would destroy almost all carbon or almost all oxygen in every star" (Barrow 2002, 155). Both elements are essential for life.

2.2.2.3 Miscellaneous Single-Parameter Examples

There are many other single-parameter cases of fine-tuning. Two stand out:

- If the electromagnetic force were 4% weaker, there would be no hydrogen. All stars would be based on helium (Section 2.2.2.1) and there would be no water (Tegmark 2003, 46).
- The cosmological constant Λ represents the expansion rate of the universe. If it were slighter greater, the expansion would increase to a point where galaxies would be unlikely. If it were slightly less, giving Λ a negative value, the Big Bang would have been followed relatively quickly by a Big Crunch. To stay within the life-permitting range, Λ cannot vary more than one part in 10^{53} (Collins 2003).[6]

2.2.2.4 Multiparameter Combinations

Some of the most interesting examples involve a combination of parameters. To consider just one, electromagnetism and the strong force jointly govern the stability of nuclei. Positively charged protons repel each other; the strong force binds them together. If the charge of an individual proton were approximately three times larger, electromagnetism would win this battle in all nuclei with an atomic number greater than five. There would be no carbon, nitrogen, or oxygen. A similar outcome would be produced if the strength of the strong force were cut in half (Barrow and Tipler 1986, 326–327).

Many of these examples of fine-tuning and more that were not mentioned govern the formation of stars:

> Were the neutron heavier by only one percent, the proton light by the same amount, the electron twice as massive, its electric charge twenty percent stronger, the neutrino as massive as the electron etc. there would be no stable nuclei at all. There would be no stars, no chemistry. The universe would be just hydrogen gas …. (Smolin 2007, 328–329)

Smolin estimates that when all of these are considered, the chance of stars existing in the universe is 1 in 10^{229}:

To illustrate how truly ridiculous this number is, we might note that the part of the universe we can see from earth contains about 10^{22} stars which together contain about 10^{80} protons and neutrons. These numbers are gigantic, but they are infinitesimal compared to 10^{229}. In my opinion, a probability this tiny is not something we can let go unexplained. Luck will certainly not do here; we need some rational explanation of how something this unlikely turned out to be the case. (Smolin 1999, 45)

It's easy to see why fine-tuning has become such a hot topic in physics. Like a person winning the state lottery 10 times in a row, the idea that we just got lucky doesn't seem all that plausible when the odds are this bad.

These examples are some of the best, in my view. Others have been intentionally passed over, such as the proton–neutron mass difference. It is often claimed that if this difference were only slightly less, then protons would decay into neutrons, thus destroying the entire periodic table. As Sober shows, this ignores further reactions that neutrons themselves can undergo (2003, 194). We would not in fact be reduced to a neutron-only universe, but one needs to take a slightly broader view of the relevant physics to see it. The point is that not every example of fine-tuning one finds in the popular literature is a good one.[7]

By some estimates, there are 50 or so examples of fine-tuning, including more local facts such as the size of our moon relative to the Earth and the solar system's place in the galaxy. Most agree that this data cries out for explanation. How all this is relevant to religion, or at least theism, is obvious. One explanation for why the universe appears to be fine-tuned is that it was. We did not win the cosmological lottery. The cosmic constants have been intentionally set to the values needed for life to exist. In what follows, we will consider alternatives to the theistic explanation. To streamline the terminology, let fine-tuned constants ('FTCs') cover both fixed parameter cases and those involving initial conditions.

Before we get to those, let's first deal with a *non sequitur*, the "well who designed God, then?" reply. The idea is that appealing to God as designer merely pushes the problem back one step and so is a waste of time. Unless theism can provide an ultimate answer to the question of design, it ought not be considered as a viable explanation. Theists have various ways of responding (see, e.g., Collins 2005), but I don't think one must have a good response in order to proceed. Whether God does or does not need a cause or explanation is an interesting question in metaphysics, but it is a different question than the one here. The issues at hand are as follows: (i) whether the universe shows evidence of fine-tuning; (ii) if so, what are the possible explanations; and (iii) among those, which is the best explanation? *If* one is granting that fine-tuning is best explained via supernatural design, then one can consider the nature of that designer. But even if the designer were completely unknown, that should not stop us from exploring (i)–(ii). The who-designed-God question is a red herring, changing the subject from fine-tuning to the metaphysics of a first cause. We need not answer that new question in order to examine fine-tuning itself. In other words, we should reject the red herring and continue on. If design is not the best explanation, we need not proceed to questions about a first cause. If design is the best explanation, then I recommend the work of a good metaphysician for sorting out the nature of the designer.[8]

When I said that "most" philosophers and physicists believe that fine-tuning cries out for explanation, I was not merely allowing room for a few crackpot skeptics. A significant, well-credentialed minority argue that this desire for explanation rests on a bad foundation. If this seems implausible— how could these coincidences *not* need an explanation!—consider that one might feel the pull an argument using an informal fallacy (e.g., *ad hominem* attacks), but that does not make the argument any less fallacious. Let's consider what the skeptics have to say.

2.3   No Explanation Needed

Three approaches have been taken to argue that fine-tuning does not need any special explanation. The first is an appeal to coincidence. The second is that the data are biased by our own observations. If the data are skewed toward fine-tuning, then they might not represent the full range of physical possibilities. The third has to do with the nature of probability itself. We will assess each of these objections once all of the problems are on the table.

2.3.1   Coincidence

An old objection to design inferences is that improbable things happen all the time. Whatever the exact number of oxygen atoms in this room at this moment, the odds of it being just that number are extremely low given all of the possibilities. Still, no one thinks we need an explanation for why that number has the precise value that it does. So long as there is enough oxygen for healthy respiration, the details don't matter. Some claim that the same goes for fine-tuning. The odds that, say, the weak force would take *any* specific value are small, given the vast range of possibilities. Why think we need an explanation for the actual value that it has?

When it comes to fine-tuning, it isn't merely the small odds involved but the extreme, negative results arising from a small change. A very slight difference in the values of the FTCs produces dramatic change in the universe when it comes to habitability. We exist in a fantastically narrow window of possibilities outside of which life is impossible. Contrast this with the oxygen-atoms-in-the-room example. We can survive with much less oxygen, with air that is polluted, with various ratios of oxygen and nitrogen, etc. Life does not depend on the precise number of oxygen atoms in the room. The amount of oxygen could change over a fairly large range but would produce few noticeable differences vis-à-vis habitability. In contrast, life itself depends on the FTCs having the precise values that they do.

One would have expected *a priori* that life is stable with respect to changes in the physical constants and initial conditions. In other words, we would expect the FTCs to behave like the oxygen example in which a slight change makes little observable difference. Small changes produce small effects. What makes the FTCs special is that slight changes in their values have effects such as altering the chemical composition of the universe! Small changes produce dramatic effects. Nonetheless, that is what current physics tells us would happen, and it is that narrow, life-supporting range in the FTCs that requires an explanation. Coincidence is not a plausible response.[9]

2.3.2 Weak Anthropic Principle

The term *weak anthropic principle* is not used in a uniform way. Let's begin with a somewhat naive view and then move on to a more significant challenge.

2.3.2.1 FTCs Are Necessary
Everyone agrees there is a sense in which the life-permitting values of FTCs are necessary. They must have the values they do; otherwise, we wouldn't exist in order to take notice:

> The basic features of the Universe, including such properties as its shape, size, age and laws of change, must be *observed* to be of a type that allows the evolution of observers, for it intelligent life did not evolve in an otherwise possible universe, it is obvious that no one would be asking the reason for the observed shape, size, age and so forth of the Universe. (Barrow and Tipler 1986, 1–2)

As we've said, the actual FTC values are necessary conditions for life itself. Most will go on to ask for an explanation, but others conclude that the questions end here. Necessary truths need no further explanation. As Richard Swinburne characterizes this line of thinking,

> We shouldn't be surprised that the values allow for life: if they didn't have such values, we wouldn't be here to pose the question! "[Unless] the universe were an orderly place, men would not be around to comment on the fact. … Hence there is nothing surprising in the fact that men find order—they could not possibly find anything else." (1979, 137)

And if there is nothing surprising here, then there is nothing to be explained. The why questions have simply come to an end.

Few skeptics put things quite this starkly anymore. It's a bit like telling a skydiver that he should not be surprised that he survived after his parachute failed. True, if he had not survived, he would not be around to wonder about it. But so what? It's ludicrous to think he shouldn't be surprised at having lived through the experience. Or consider Sleeping Beauty. Let's say that before she falls to sleep, she comes to know that she has been cursed and that only the kiss of a prince can awaken her. Let's amend the story so that the witch intends to transport Sleeping Beauty to the *Star Trek* world of Delta Vega on which there are no princes. (It's *my* amendment; I can do what I want.) Sleeping Beauty now realizes that the odds of the curse being broken are tiny indeed. Say that, just like the story, she wakes up to the kiss of a handsome prince. Should she be surprised? Not according to the weak anthropic principle. If a prince had not found her, she never would have awoken. The kiss was a necessary condition for her to even notice that she was awake, and so there is no need for further explanation. I find that conclusion implausible and her request for an explanation to be completely reasonable. The same goes for fine-tuning.

There is still work to be done, however, as there is a more rigorous way to understand the weak anthropic principle.

2.3.2.2   *Observational Selection Effect*

Consider a well-known analogy (Sober 2009, 77). Say you use a net to capture some fish in a lake. All of the fish you net are over 10 inches long. This would seem to favor the hypothesis h_{all} = 'all of the fish in the lake are over 10 inches long' over hypothesis h_{50} = '50% of the fish in the lake are over 10 inches long.' But what if you find that the mesh of your net only has 10 inch holes? You couldn't possibly catch a fish under 10 inches, and so the observation was inevitable. In this case, you don't know whether h_{all} or h_{50} is more likely to be the case. The observation gives you no relevant information about the fish population. The size of the net determined the observation, not the size of the fish in the lake.

When it comes to fine-tuning, Sober considers h_{INT} = 'the constants have been set in place by an intelligence, specifically God,' and h_{Chance} = 'the constants are what they are as a matter of mindless random chance.' Sober argues that just as the net can only catch big fish, you, the observer, can only observe a universe with FTCs in the life-permitting range; otherwise, you wouldn't be alive. In other words, you are playing the role of the net in this example. The net can only capture big fish; you can only observe life-permitting FTCs. But just as h_{all} could not be confirmed over h_{50}, in this case, h_{INT} cannot be confirmed over h_{Chance}. An observational selection effect blocks h_{INT} from raising the likelihood of fine-tuning above that of h_{Chance}.

Sober's argument, which he casts more rigorously in terms of Bayesian probabilities, continues to be a matter of debate.[10] A crucial disagreement is whether your existence should be treated as (i) merely a possible outcome or (ii) a known fact since, after all, you do exist. Collins argues that in the case of fine-tuning, your existence is the one thing that must *not* be included in your background knowledge, directly contradicting Sober (Collins 2009, 241). Who is right depends on what one does with "old evidence"—a notoriously difficult problem in Bayesian probability theory.[11]

If this is the crux of the disagreement, as it seems to be, then it will be very difficult to resolve to anyone's satisfaction. Arguments over what to include in one's background knowledge devolve into conflicting intuitions, especially when the very issue at hand is observer selection effects. Once the debate gets down to dueling probability theorists complaining about what the other guy includes in his background knowledge, I start looking for a way out.[12] What we need is a different approach to framing the debate.

Let's think of it in terms of what does and does not require an explanation. Sober doesn't say so explicitly, but Chance must represent a physically random event such as symmetry breaking (Section 2.2.2). An appeal to Chance means that no further explanation is needed. Whatever the mechanism, we essentially won the cosmological lottery and the FTCs fell in our favor. Consider another random event, say, drawing the Jack of Hearts from a deck of playing cards. Surprising? No, again, this is just a matter of Chance, a random event where the probability is 1/52. No special explanation is needed when an event is due to Chance. It just happened by way of physically random processes (Sober 2003, 31–32). In Sober's argument, he likewise shows that Chance is as good as any rival explanation. Philosophers of science would say that Sober has "explained away" fine-tuning. Because of selection effects, it really isn't a problem in need of a concrete explanation.

If so, then supernatural intelligence is not the only enemy at Sober's gate. His argument is equally damning for all those who believe that fine-tuning cries out for an explanation. One could replace h_{INT} with any of the proposed naturalistic explanations of fine-tuning that we will examine later in this chapter, and the whole argument goes through as before. A large number of physicists are currently wasting their time since, if Sober is right, there are *no* cases of fine-tuning in need of explanation. All the fine-tuning data are the result of selection effects (Sober 2003, 53n21). I will say more about why I think that's a problem at the end of this section. Let's now turn to objections to fine-tuning based on the nature of probability itself.

2.3.3   The Nature of Probability

This section gets technical and some readers may wish to skip it. On the other hand, this is one of the most serious objections raised by those who know the topic well. It is seldom covered in the more popular literature.

In cases with discrete, finite numbers of equally possible outcomes, probability calculations are easy: Pr(fair die coming up 5) = 1/6, Pr(heads on flipped coin) = 1/2. But what if things are not equal, say, with a loaded die? Now, the right probabilities should reflect that there is a greater chance of coming up a certain value, say, Pr(die coming up 5) = 2/6, and the other five reduced according. However this redistribution goes, all of the probabilities together must equal 1 (i.e., 100% probable).

Things are still more complicated when the possible outcomes are not discrete. For example, what are the odds of hitting a bull's-eye on a target? Now, we have to account for the relative area of each part of the target. This is also the case when talking about probability and abstract spaces, like the state spaces mentioned earlier (Section 2.2.1). In order to calculate probabilities, we first have to find the *measure*, μ, of the space. The concept of measure is a generalization of length, area, and volume. Given a continuous n-dimensional space of possibilities, where n is any positive integer, the measure indicates relative "volumes" of different parts of the space. The simplest way to get probabilities from measures is to take them as roughly equivalent. The larger the measure of a given area, the higher the probability that a system state will be in that area. If you think of a probability distribution as paint, then big areas get more paint proportionally.

Here's one more example that we will need to refer to later. *Phase spaces* are used in dynamics to represent the state of a system and its evolution. Consider an ideal pendulum. Much like Cartesian coordinates on a flat surface, every possible state of the pendulum can be represented by two numbers. Instead of (x, y), the numbers in this case are the angle of the pendulum θ and the angular velocity of the swinging bob, $\dot{\theta}$, multiplied by a constant. One point in the phase space represents one system state $(\theta_1, k\dot{\theta}_1)$. As the state evolves over time, a trajectory is carved through the space. Every point belongs to some possible trajectory which represents the evolution of the system over time.

Figure 2.4 shows one such trajectory.

Now, what is the probability that, at a random instant in time, we will find the state of the pendulum around the extreme right side of this orbit, close to $(\theta_{max}, 0)$? Once again, we will have to find a measure for this space. With a measure μ and a probability distribution ρ in hand, one can integrate over an area A to find the probability of finding the state point in A[13]:

$$Pr(A) = \int_A \rho\left(\theta, k\dot{\theta}\right) d\mu$$

Finding the right measure of a physical space, like a dart board, is usually not a problem. Finding proper measures for some mathematical spaces, like phase spaces in classical mechanics, is also relatively easy. (Because of the mathematical niceties of Hamiltonian mechanics, a so-called "natural measure" presents itself.) In other types of systems where the space is more complex, there is often no clear way to determine the measure. This is where the problems start.

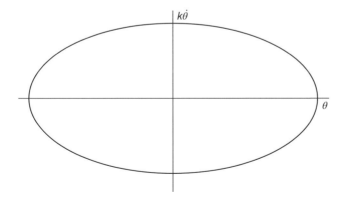

Figure 2.4   Ideal pendulum phase space.

2.3.3.1   There Can Be No Measure

Philosophers Timothy and Lydia McGrew and mathematician Eric Vestrup argue that when it comes to FTCs, probability has no meaning (2001). Let *A* stand for one of our FTCs. Say that *A* could have taken a value from 0 to any positive real number (since most FTCs have no theoretical bound). We may think of this range as marking off a space of physically possible worlds, that is, ones that have the same laws of nature as our universe but differ with respect to *A*. More precisely, this is a 1-dimensional coordinate space for the possible values of *A* in the interval $[0, \infty]$. Since nature, as far as we know, has no preference for one value of *A* rather than another, all subintervals of equal size should be assigned equal probability. Life-permitting values of *A* are restricted to a subinterval with a tiny measure compared to the full measure of this space. Hence, the argument goes, the odds of the universe having a life-friendly *A* region is small—the proverbial needle in an infinite haystack. The fact that our universe contains life therefore requires an explanation.

Very well, except for one thing: haystacks are not infinitely large. Does it make sense, even as a mere thought experiment, to talk about the *probability* of drawing a specific ball from an infinitely large urn? Many probability theorists will reply with a resounding "No!" Probability distributions cannot be defined in these circumstances. When one sums up each bit of a uniform probability distribution over a space of infinite measure, the total is infinity. That's bad. Probabilities must add up to exactly one. Infinite spaces break the rules:

> This is more than a bit of mathematical esoterica. Probability has no meaning unless the sum of the logically possible disjoint alternatives adds up to one. There must be some sense in which all the possibilities can be put together to make up one hundred percent of the probability space. But if we carve up an infinite space into equal, finite-sized regions, there will be infinitely many of them. And if we try to assign them each some probability, however small, the sum of these is infinite. (McGrew *et al.* 2001, 1030)

So although it might seem that one can make a probabilistic argument about fine-tuning, such thinking does not apply to infinite spaces. The upshot is that "there is no meaningful way in such a space to represent the claim that one sort of universe is more probable than another" (McGrew *et al.* 2001, 1032). There is strictly speaking no sense in which life-friendly universes are improbable. The probabilities are not defined.

Even if we can solve this problem, as I believe we can, a closely related one must be addressed.

2.3.3.2 There Is No Accepted Measure in Cosmology

Regardless of whether a measure might possibly be defined on the relevant spaces, the fact is that no one has such a measure in hand when it comes to FTCs. Without one, argues Callender, we can't know what does and does not need explaining (2004, 200–203). Callender's objection focuses on one specific case of fine-tuning, the low-entropy initial conditions, but it applies to others.[14] We said earlier that the measure for some spaces is obvious. But what if the "right" variables aren't so clear (i.e., there is no natural measure)? Say we want to know how fast a curved soda pop is being filled (Sklar 1993, 119). We have a choice to make. Do we measure the rate according to the volume of liquid inside or to the surface area of the bottle that is darkened by its contents? While both are perfectly good physical parameters, the interior surface area is not directly proportional to its volume in such bottles. Hence, the answer to "how much soda pop is in the bottle?" will vary depending on which parameter one chooses. A measure of the space in terms of volume will not match a measure in terms of surface area. So then, which should the rational person choose? There is no answer; the notion of uniform measure for liquid in such a bottle is not well posed.

At least with the bottle we have complete access to the system in question. Say you walk into a shop and see a large deck of unusual cards. All the cards are facedown. You pick one up at random, and all that is printed on the other side is the number 7, nothing else. The next day, you do the same thing: pick one card, it's a 7. You do this for 100 days with the same outcome each day. Is this surprising? It would be if this were a normal deck of cards since there are only four 7s in the deck. But you don't know anything about these particular cards other than what I've told you. Perhaps they're all 7s. That wouldn't make for much of a game, but who said this was for a game? Perhaps the deck was designed by someone who likes 7s.

To know whether the data are surprising and require an explanation, you need to have more information. This information would allow you to construct a rough probability measure on the deck. If you knew what each card was, we could do so precisely. Without this information, you have no idea whether the phenomena are surprising or not. The same goes for cosmology (Manson 2000, 348). The standard FLRW models lack a natural measure. Without one, probabilistic arguments have no foundation, and we therefore have no proper basis for what is and is not surprising, including fine-tuning.

This is just a technical way of saying that whether one is surprised by an event requires some background knowledge about that event. Say I roll a six-sided die 20 times and that 19 of those times, it rolls a one. Should I be surprised? If the die is fair, yes. If the die is loaded, no. In cosmology, we don't know if the results are loaded, that is, whether there is a physical bias toward the kind of universe we have. And unlike the dice example, you can't do multiple trials in order to determine the bias. We only have this universe to work with. Without a natural measure—without a reason to believe we're thinking about the range possibilities in the *right* way—we cannot make valid inferences about what is or is not surprising. Ignoring this fact distorts the scientific process. Commenting on one proposed explanation for fine-tuning, Callender says, "It posits a substantive—and enormously extravagant—claim about the world in order to satisfy an explanatory itch that does not demand scratching" (2004, 213). I'm sure he would say the same about supernatural design.

As if that weren't enough, there is one more challenge to the idea that fine-tuning needs an explanation.

2.3.3.3   Coarse-Tuning

The power of fine-tuning comes from the small life-permitting ranges that FTCs can take. As we have seen, in many cases, there does not seem to be any bound for how large a given parameter might be. If the value of a parameter A can take any value from $[0, \infty]$, the parameter space will be infinitely large. In terms of measure, however, this presents a problem. The measure of the life-permitting range of A is zero if the background parameter space is infinitely large. If we allow measure to serve as a surrogate for probability, then the $Pr(A$ taking a life-permitting value| the parameter space for A is infinite) $= 0$. Normally, that would strengthen the need for an explanation. If some event appears to have an infinitesimal probability, and yet it happens, such an event would need an explanation. *Something* must be going on, even if it isn't clear what. So if A is within a life-permitting range (we exist), and yet the physics tells us that the probability of A is zero, there must be more to the story. We need an explanation over and above what the physics is telling us thus far.

But there is something else to notice here. For $Pr(A) = 0$, only two things are needed: the parameter space is infinite and the life-permitting range for A is finite. Any fixed interval for A, no matter how large, will have zero measure in the space of positive real numbers. Hence, if the background space is infinite, the so-called fine-tuning is virtually guaranteed (Manson 2000, 347). One would get the same measure-theoretic results if the range of each fine-tuned parameter were "within a few billion orders of magnitude" of their actual values (McGrew *et al.* 2001, 1032). Presumably, no one would say that coarse-tuned parameters cry out for explanation. After all, the motivation for the original argument is that so many FTCs are seemingly balanced on a knife-edge. A slight change entails that life is impossible. If life were still possible even when the parameters were wildly different, there would be no surprise and nothing to explain.

The problem is that in measure-theoretic terms, fine-tuning and coarse-tuning are equivalent. If coarse-tuning does not require an explanation, then fine-tuning ought not either.

We have just examined three arguments based on probability and measure. Each concludes that, contrary to what seems to be the case, fine-tuning does not in fact require an explanation. For my part, I believe that these objections can be rebutted and that the majority view on this question is correct: fine-tuning does need some sort of explanation. Let's now consider some replies aimed at defending this view.

2.3.4   Analysis and Replies

I'll address each of the three arguments just made and then give a more general reply to arguments of this sort at the end of this section. A preliminary note is that there are many interpretations of probability (see Sklar 1993, 96–120), and I do not think that any one is *the* correct view. The word 'probability' is ambiguous, and we use different types in different situations. However, I tend to favor the objective interpretations over epistemic or subjective ones whenever it is appropriate, including what follows. Intuitively, I take it that probabilities are for the most part "out there" to be discovered rather than degrees of belief in us.

2.3.4.1   *The Problem of Infinite Measures and Defining Probabilities*
McGrew *et al.* argue that probabilities are undefined across a space of infinite measure. In such a case, the probabilities of each subinterval add up to infinity rather than 1.

One simple solution is to truncate the range. Even though the entire mathematical space in question is infinite, one might limit it to some arbitrarily large, finite number. In other words, even if the physically possible range of A is $[0, \infty]$, the interval $[0, N]$ works just as well for our purposes, where N is a very large number relative to A. The space is now finite and probabilities can be defined as usual.

"Sure," the skeptics will reply, "but that's just cheating. Do you think you can just ignore mathematically inconvenient possibilities?" In short, yes. Physicists idealize and simplify matters all the time. Changing large numbers into infinite ones and vice versa are common techniques for making incalcitrant mathematics more manageable. Usually, there is some physical justification for doing so. When it comes to the force coupling parameters, truncating the ranges is grounded in the fact that their values place bounds on one another. It appears the fundamental forces can only take on a limited range of strengths relative to each other. Once one parameter is fixed, the range of the others would likely be limited, and so there cannot be infinite measures for each of the four coupling constants.

In other cases, there is no physical difference after an FTC reaches a sufficiently high value. Still, higher values would produce the same outcome. For example, there is a point at which the strength of gravity would be so strong that the universe would have collapsed back on itself before stars could evolve after the Big Bang. Once the collapse occurred, the universe would be nothing more than a high-entropy singularity, like a black hole. For any higher value of the gravitational constant beyond this point, the result would be the same. Hence, one could set the parameter space for G from $[0, N]$ where N is well into the range where this universal collapse would occur. The parameter space is now finite, and again, probabilities can be defined as usual.

All this mitigates the problem of infinite measures but does not solve it. What about those cases where there is no good physical reason for truncating the space? This was precisely the issue for cosmologists in the 1980s. Most physicists believe that the early universe underwent a temporary, exponential expansion known as *inflation*. The stretching of space caused by inflation is supposed to explain, among other things, why space is very nearly flat (Euclidean). In order to determine whether inflation was to be expected in a universe like ours, Gibbons *et al.* (1987) derived a measure across the space of Big Bang models. Since, as far as we know, inflation is no more likely in one set of FLRW models than another, they assumed a uniform probability: intervals of equal measure are equally probable. They were attempting to show that almost all of the models within the space undergo inflation. *Almost all* is a technical term here: "all but a set of measure zero when the measure of the entire space of possibilities is finite, or for all but a set of finite measure when the measure of the entire space of possibilities is not finite" (Earman and Mosterin 1999, 31). If *almost all* of the points in a given space have some property Q, then $Pr(\sim Q) = 0$ for that space. Q is what's normal, what one would expect. If black holes were to occur in almost all of the FLRW models, then black holes are the norm. If we lived in a universe relevantly similar to an FLRW universe but there were no black holes, *that* would require an explanation.

Note that *almost all* is defined in spaces of infinite measure, although the terminology is not uniform:

> Indeed one popular way of explaining cosmological observations is to exhibit a wide class of models in which that sort of observation is "generic" [i.e., almost all of the models have it]. Conversely, observations which are not generic are felt to require some special explanation, a reason why the required initial conditions were favoured over some other set of initial conditions. (Gibbons *et al.* 1987, 736)

Here, *generic* is equivalent to *almost all*. (In nonlinear dynamics, *prevalent* is equivalent to *almost all*; *generic* is slightly weaker.) Let's see how this works. The geometry of space is surprisingly Euclidean or "flat" given the vast range of non-Euclidean possibilities. Most physicists believe that flatness needs an explanation; inflation is the current favorite.[15] Skipping all the details, let's assume that inflation would account for flatness and that inflation occurs in almost all FLRW models (i.e., except for a set of finite measure within an infinite model space). In such a case, flatness would not require any further explanation.[16] If inflation produces a flat spatial geometry and inflation is the norm—what the physics is expected to produce—nothing more is needed. This is precisely what *almost all* arguments are supposed to accomplish.[17] If our universe did not experience inflation and were not flat, *that* would require an explanation.

The key point here is that these are inferences made on spaces with infinite measure. Contrary to what the critics suggest, information about probabilities can be derived in such cases.

Some will complain about the lack of rigor here. Gibbons, Hawking, and the rest might refer to "probability measures" on infinite spaces, but that is just the sort of mathematical hand-waving that the critics are complaining about here. If one is going to make probabilistic arguments, they argue, one must play by the rules.

While we should acknowledge the point, debates about rigor are nothing new. Mathematicians and physicists have been arguing about such things for centuries. As philosopher Mark Wilson puts it, "Physicists commonly employ inferential principles in circumstances that the applied mathematician cannot semantically underwrite" (2010, 559). Nonetheless, advances in both physics and engineering would have been disallowed if every move had to pass the mathematician's test for rigor. Consider the case of Oliver Heaviside (1850–1925), one of Wilson's favorite examples. Heaviside pioneered a method for analyzing electrical circuits that we now call Laplace transforms. At the time, however, his techniques were so wildly unorthodox that the Royal Society refused to publish them even after Heaviside had become a fellow of the society. No one questioned that Heaviside was able to use his "operational calculus" to get the right answers. Their complaint was that "the rigorous logic of the matter is not plain!" Consider his reply:

> Well, what of that? Shall I refuse my dinner because I do not fully understand the process of digestion? No, not if I am satisfied with the result. Now a physicist may in like manner employ unrigorous processes with satisfaction and usefulness if he, by the application of tests, satisfies himself of the accuracy of his results. At the same time he … may be repellent to unsympathetically constituted mathematicians accustomed to a different kind of work. (Heaviside 1894, 2:9)

The unaccommodating Heaviside was not pleased to have his discoveries dismissed. His methods worked regardless of whether "Cambridge mathematicians" could fully make sense of them. "Objections founded upon want of rigour seem to be narrow-minded, and are not important, unless passive indifference should be replaced by active obstructiveness" (Heaviside 1894, 2:220–221).

I believe this is one of those cases. We can all agree that cosmologists stretch orthodox probability theory. Mathematicians can demur, but that should not stop scientists from using what works. Specifically, inferences about what does and does not need an explanation can be made on spaces of infinite measure.

This might be a small victory, however. *Almost all* reasoning in spaces with infinite measure leads directly to coarse-tuning: if the parameter space is infinitely large, then even vast ranges of life-permitting values will still have measure zero in that space. But if the FTCs could be very different and still allow for life, *that* wouldn't have surprised anyone. Coarse-tuning would not seem to need some further explanation.

2.3.4.2   *Coarse-Tuning and Measure Zero*
I think we can agree that if the FTCs were coarse-tuned rather than fine-tuned, few would have found this surprising. Books would not have been written, and research would have focused on other matters. Coarse-tuning

would have gone unnoticed. That is not the end of the story, however. There are three points to be made in response.

First, keep in mind that the coarse-tuning objection applies to cases where the parameter space is infinitely large. Although fewer researchers would have noticed, I believe that coarse-tuning would still require an explanation. Recall the quote by philosopher of physics John Earman defining *almost all*: "all but a set of measure zero when the measure of the entire space of possibilities is finite, or for all but a set of finite measure when the measure of the entire space of possibilities is not finite" (Earman and Mosterin 1999, 31). *Almost all* reasoning is valid for both finite and infinite sets. If one can show that almost all of the points in a given space lack some property Q, then $Pr(Q) = 0$. If we find that Q is actually the case, that would require a special explanation. The critics hope to use coarse-tuning to show that if coarse-tuning does not require an explanation, then neither does fine-tuning. I claim that the antecedent is false. If the life-permitting constants have measure zero in the relevant space, that requires an explanation regardless of whether those constants are fine- or coarse-tuned. Fine-tuning advocates ought to bite the bullet, trust the mathematics, and say that a coarse-tuned universe *would* require an explanation.

Second, critics use coarse-tuning to aim at a specific target, namely, FTCs in cosmology. But fine-tuning is just a special case of *almost all* reasoning which has applications in several areas of physics. Their beef isn't with fine-tuning so much as measure-theoretic inferences on infinite spaces. This multiplies the enemies at the gate. If the coarse-tuning objection is correct, then a good deal of research in mathematical physics from the last 30 years has been a waste of time, especially in statistical mechanics and chaos theory.[18] That, it seems to me, is a much larger foe than fine-tuning critics were hoping to take on. It also shows that the objection is too strong. I see no way of using coarse-tuning as narrowly as the critics had intended.

Third, there is one thing right about the coarse-tuning objection. Those fantastically narrow ranges of the FTCs are what get one's attention, no doubt. Coarse-tuning within an infinite parameter space might not have gotten much notice. That, however, points to a limitation on our cognitive abilities, not on the strength of the argument. The fact remains that a life-friendly, coarse-tuned universe is part of a set with measure zero. Again, if $Pr(Q) = 0$ and yet Q is the case, that requires an explanation even if most people would fail to appreciate it. After all, few people know enough about measure theory to properly frame the issue in the first place. So what? Only a small percentage of people in the world can appreciate the difficulty of

Goldbach's conjecture,[19] but this is a matter of education and intellect, not the force of Goldbach's challenge. The same would be true if the universe were coarse-tuned for life, assuming as the objection does that the parameter spaces are infinitely large. Fine or coarse pertains to how obvious the anomaly is, not whether there is an anomaly in need of explanation.

2.3.4.3 *Arbitrary Measure*

Even if cosmological measures exist as matters of pure mathematics, the last issue we must face is finding the right one. Knowing that an acceptable and precise answer to a question exists is not the same as knowing the answer itself. (How many water molecules are in my coffee mug?) The critics charge that there is no nonarbitrary measure for use in cosmology. Although a few published articles suggest otherwise (e.g., Evrard and Coles 1995; Kirchner and Ellis 2003), let's nonetheless grant that these are special cases and there are no natural measures in cosmology applicable to most FTCs. What should we say? Some argue that securing a measure must be the first step. Only then can we know if there is an anomaly that needs to be explained. "Debates about likely versus unlikely initial conditions without a well-defined probability are just intuition-mongering" (Callender 2004, 202).

In my view, Callender demands too much. A more typical approach is taken by Kirchner and Ellis (2003). Although they do not claim to know the right measure across model universes, they do show that there is a measure that solves the flatness problem. A flat spatial geometry is overwhelmingly likely in this measure, it turns out, and so there is no need for a mechanism like inflation to produce it. If something is likely to happen on its own account, nothing more is needed. However, they acknowledge that their measure is limited (only over FLRW dust models) and speculative (new information could show it's wrong). What they do not say is that physicists should wait until they know the right measure before doing more research. Kirchner and Ellis show how the flatness problem *might* solve itself. That's very different from saying that we don't even know if there is a problem to be solved until we have the right measure, which seems to be Callender's view.

Here is the situation with flatness in slightly different terms. The data seems to be anomalous and as such requires an explanation. Inflation is one popular way of accounting for this anomaly. Kirchner and Ellis show that, *pace* the majority, flatness might not need a special explanation after all. Instead of being anomalous, it might in fact be overwhelming *likely* in the space of possible universes—something to be expected rather than a surprise in need of explanation. The jury is still out. Their measure might not properly capture the underlying physics, much like a uniform probability measure applied to a loaded die would fail to capture the underlying physical bias of that die. For now, research continues on all fronts, and it isn't clear where the right answer about flatness lies.

Notice that no one—Kirchner and Ellis or the inflation theorists—is waiting around until someone finds the correct measure. That, however, seems to be what the "no measure" objection demands. Callender and Manson want a natural measure to justify the sense that something needs to be explained. "After all," they would say, "the anomaly might not turn out to be anomalous. It might turn out to be what we should expect." Well, it might, but the issue is whether the data need an explanation given the totality of what we know today. We need not have the right measure in hand to justify the demand for explanation. Requiring the natural measure first is too great a burden on research.

2.3.4.4 Catch-All Response

This leads to my final, catch-all response to skeptical arguments based on probability, measure, and selection effects. Keep in mind that there are two questions here: (1) does fine-tuning need an explanation, and (2) if so, then what is the best explanation? Each of the objections in this subsection answers 'no' to the first question, at least with the available evidence we have in hand. If Sober is correct, then fine-tuning is the result of an observational selection effect just like netting the lake in the fish. Once you understand things properly, the data that seemed to require an explanation in fact do not. Others say that the underlying mathematics undermines the need for explanation. In any case, fine-tuning arguments are either misguided or at best premature.

If any one of these is right, then hundreds of world's best physicists have engaged in decades of work for nothing. Could Hawking, Penrose, Susskind, Linde, and Guth all be wrong, and in fact, fine-tuning is not in need of explanation? Of course. A thousand years ago, every educated person in the world believed that the sun revolved around the Earth. But whose intuitions should we trust at this stage of the game? Note, I'm asking about question (i), not (ii). Physicists overwhelmingly agree that the answer to that question is yes. They are split regarding the second, as we will see. My argument is not merely an appeal to authority: "All these smart physicists think that fine-tuning needs to be explained, therefore it does." Rather, the opinion of experts establishes the burden of proof.[20] Fine-tuning skeptics have to do more than merely undermine our sense of surprise; they need an argument that tips the balance in their favor. Given the flaws in those arguments, I am strongly inclined to side with the physicists on this one. Discoveries from the last half of the twentieth century *are* surprising, and fine-tuning does require an explanation, even though we might disagree about which explanation is best.

Until the skeptics meet this burden, research on these questions ought to continue. Inflation theorists are completely within their epistemic rights to see flatness and other phenomena as data requiring an explanation. Physicists do not need—*rationally* not need—a measure in hand first. At most, the skeptical arguments show that if one is inclined to write off fine-tuning as a matter of chance, that position is not completely baseless.

The objections in this section were mostly of interest to philosophers. Physicists overwhelmingly see the fine-tuning data as something in need of an explanation; however, most do not believe that a theistic explanation is a scientific one. Let's now consider some of the alternatives. If not design, then what?

2.4   Naturalistic Explanations

Many naturalistic explanations have been offered for fine-tuning. Here, we consider the three most important ones. The first is that future discoveries might show that fine-tuning was not improbable after all. The second is that other life-permitting ranges for the FTCs might exist, allowing for exotic forms of life unlike ours. The third and most popular is that we might live in a multiverse of many different universes all with different values of FTCs. On that view, we merely inhabit the corner of the multiverse that happens to support life. Each of these proposals will be critiqued in turn.

2.4.1   Future Physics

At present, the actual values of the FTCs must be determined by experiment. So far as we can tell, their values are brute facts: there is no physical reason why they are what they are. But what if this only reflects the state of current physics? Just as inflation came to solve the flatness problem,[21] new discoveries might account for the FTCs (McMullin 2008, 77). A yet more fundamental law of nature or physical process not yet discovered might show that these constants *must* take the values that they have. If so, then the observed values of the FTCs are not highly unlikely. There are what they must be as a matter of natural law.

Let's consider the best candidate for a law in which FTCs are not free parameters. *Supersymmetry* is an idea based on unification (Smolin 2006, 66–79). We've already discussed how electromagnetism and the weak nuclear force are low-energy manifestations of a unified electroweak force. It is widely believed that this unification of forces can be pushed further at still higher energies. The so-called "grand unified theories" (GUTs) look for ways of folding the strong nuclear force into the electroweak force. If gravity could somehow be incorporated at still higher energies, then all four forces would be unified. The forces we observe would merely be low-energy manifestations of this one fundamental force.

Even if this unification is possible, the division between forces and particles would remain. Forces are mediated by force carriers (bosons) such as the photon. Particles (fermions) include the more familiar electron, proton, neutron, and neutrino. Supersymmetry is the idea that that bosons and fermions might somehow be combined into one basic kind of stuff. At the most fundamental level, there would be no distinction between particles and forces.

In order to pull this off, supersymmetry would require an array of new particles that are counterparts to the ones we already know. For example, for each quark, a fermion, there should be a squark, its boson "superpartner." The photon is paired with a new fermion, the photino, and so on down the line for each fermion and boson. In the end, somewhere between 20 and 125 new parameters would need to be introduced. While we've never observed these new particles, they are assumed to exist at extremely high energies which the Large Hadron Collider in Europe might reach.

One might wonder why physicists would consider adding so much structure to the already complicated Standard Model. The answer is that these new parameters might not be fine-tuned. By adding new parameters, the hope is that the FTCs will turn out to be a consequence of a theory that is not fine-tuned. If so, then small changes in the new supersymmetry parameters would not lead to large differences in the nature of the universe.

So how likely is it that supersymmetry will solve fine-tuning? I've skipped all the details, but the answer is, "it's possible." More degrees of freedom and massive complexity might just be the price for a solution. What seems more likely at this point, however, is that it just pushes the problem back one step. Fine-tuning is a bit like whac-a-mole: you can knock it down in one place only to have it pop up in another. Early versions of inflation faced this. While it could explain the fine-tuning associated with flatness and isotropy, inflation itself required fine-tuning in order to achieve a graceful exit.[22] So even if supersymmetry is able to explain a number of known FTCs, which isn't yet clear, it might carry with it even more fine-tuning among its many new parameters. Fine-tuning is easy to displace but hard to eliminate. As Smolin sums up, "Despite all the progress in gauge theories, quantum gravity, string theory etc. not one of these problems has been solved. Not one mass or coupling constant of any particle considered now to be elementary has ever been explained by fundamental theory" (Smolin 2007, 329).

One can always bet that future physics will solve fine-tuning, but a bet is not an argument. Given the overwhelming evidence for fine-tuning, one might rightly demand more than a promissory note.

2.4.2   Other Forms of Life Possible

This is always the first student objection to fine-tuning. If *Star Trek* has taught us anything, it's that life can take many different forms.[23] Just because the kinds of life we are familiar with could not exist if the FTCs took on different values, that doesn't mean that other forms of life are impossible. Exotic, alien creatures might thrive in a universe that we would find uninhabitable. Fine-tuning is therefore based on a kind of speciesism, as if we are the only type of life worth considering.

Although seriously flawed, this idea is so intuitive that is it difficult to pry loose. The first thing to note is that a basic requirement for life is some mechanism for overcoming the second law of thermodynamics. Without the ability to absorb low-entropy forms of energy and get rid of high-entropy waste, no life-form can exist for long. Even exotic kinds of life must have something analogous to metabolism and respiration that take in energy from outside the entity. Hence, all life requires a minimum amount of complexity. Any sort of entropy-reducing mechanism will have some differentiated structure to it. Homogeneity and life are incompatible.

Unfortunately for our student objector, such a structure is only possible within the life-permitting range of the FTCs. Many of the examples at the beginning of this chapter had to do with the creation of stars, the main source of usable energy in the universe. As I mentioned earlier, stars also produce most of the elements on the periodic table. Without stars, the universe would lack the basic chemistry needed to evolve any sort of metabolic structure. A universe of free-floating atoms cannot support life except in science fiction.

If this weren't enough, some fine-tuning examples allow for no inhabitable universe whatsoever outside of the life-permitting range. Consider this one from physicist Stephen Barr:

> [If] the cosmological constant Λ were not of order 10^{-120}, but of order 1, then either the universe would last only 10^{-43} seconds (if Λ were negative), which is too short for anything to happen, let alone for life to evolve, or the universe would double in size every 10^{-43} seconds (if Λ were positive), which would rip apart any structures, even atoms and atomic nuclei. In fact, to avoid atoms being ripped apart (if $\Lambda > 0$) or to have the universe last even for the length of time it takes an electron to go once around an atomic nucleus (if $\Lambda < 0$), $|\Lambda|$ has to be smaller than about 10^{-54}. (Barr 2010, 916–917)

A slight change in the cosmological constant would produce either a Big Crunch singularity or a universe devoid of atoms. Either way, life—any sort of life—would be physically impossible. In short, the appeal to other possible types of life ignores that a universe with any discernible structure depends on fine-tuning.

2.4.3   The Multiverse Reply

The most important naturalistic explanation proposed for fine-tuning is that this universe is not the only one. Our observable universe is one of many, perhaps infinitely many, all within a massive *multiverse*. Instead of one chance to get the FTCs to line up just so, nature has had many chances. Each universe has its own values for the constants and initial conditions we have discussed, and so most of the multiverse will be lifeless. On this view, at least one universe, our own, had all the right values and so here we are. While it might be unusual to get two royal flushes in a row, such an event is inevitable if one plays in trillions of poker tournaments. The same goes for life in the multiverse.

Three kinds of multiverses have been proposed. Let's consider these in turn and then evaluate them in the next section.

2.4.3.1   The Serial Multiverse: One Universe after Another
The idea is simple. Instead of one Big Bang, there have been many, forming a long sequence of expansion, contraction, expansion, and so on. The multiverse is created one universe at a time over many aeons. Every once in a great while, one is produced in which all of the physical constants permit life.

While intuitive, many types of serial multiverse are precluded by the Hawking–Penrose singularity theorems. According to the general theory of relativity (GTR), the size of the universe at the Big Bang was infinitesimal: there was no spacetime. In other words, the Big Bang is a true singularity, not merely a contraction to a small size and reexpansion. As such, it cannot be a physical porthole to some previous universe.

Many ways of getting around this obstacle have been proposed, the most popular of which is to reject GTR (Craig and Sinclair 2009). This is not as radical as it sounds. Relativity is a classical theory, but quantum effects cannot be ignored immediately after the Big Bang. The universe was extremely dense at that time, and so gravity was also important, but not the kind of gravity used in classical GTR. Some form of *quantum* gravity is needed. Two ideas have received a lot of attention although neither is widely held according to philosophers Hans Halvorson and Helge Kragh (2013).

The first is the Steinhardt/Turok Ekpyrotic model (http://wwwphy.princ eton.edu/~steinh/npr/). Borrowing an idea from string theory, this model proposes that our 3-dimensional space (or "brane") is not alone. There is another 3-dimensional universe coupled to our own in a springlike fashion. Once on the order of every trillion years, these universes collide producing what we call the Big Bang. The kinetic energy of this collision is converted into mass (fermions) which eventually coalesces into stars, planets, and the rest. The idea is that this clash of universes has happened many times; some claim an infinite number of times, although 'infinite' is being used here rather loosely (Craig and Sinclair 2009, 168–169).

The second is Penrose's conformal cyclic cosmology (2006). The big problem with a repeated cycle of universes is entropy. Just after the Big Bang, entropy was very small. As usable energy is expended, entropy will in time become extremely large. For there to be another Big Bang in the very distant future, there needs to be a mechanism for "resetting" the universe— putting it back into a low-entropy state. Once a battery runs out, it doesn't spontaneously recharge. Penrose's solution is to treat black holes as entropy eaters. It is widely believed that black holes have a fixed lifespan. Once they evaporate—a long, slow process—they take entropy with them, essentially recharging the battery of the universe.[24] Through the mathematical magic of "conformal rescaling," the state of the distant future can be made to look more or less the same as the distant past ("look" in the sense that the mathematical state descriptions are roughly equivalent). Penrose then connects the end of one "aeon" to the beginning of another: our extreme future will lead to another Big Bang for the next universe in the series. Before our

Big Bang, there was a previous aeon, and so on as far back as one can imagine. The ubiquity of black holes even in universes with different values for
the FTCs ensures that entropy is reduced at the end of each aeon.

The idea of serial, single universes is not the majority view, however, even
among multiverse enthusiasts. Let's consider two others.

2.4.3.2 Single Universe with Many Domains

A more popular view is that our observable universe is just one domain
within a large and complex spacetime structure. Each domain has its own
values for the FTCs. A well-known mechanism for generating these
domains is Andrei Linde's *eternal inflation* (1994). It begins with the posit[25]
that the universe is pervaded by a (scalar) energy field, the *inflaton field*,
which has different values in different regions, all of which are subject to
quantum fluctuations. Another well-established idea is that in GTR,
pressure is itself a source of gravity, usually too small to be noticed. (The
pressure at the center of the Earth is only a small fraction of the gravity we
feel.) The inflaton field has negative pressure (think tension) and therefore
opposes gravity.[26] Regions with large inflaton field values inflate under the
influence of this negative pressure. (Since the values are randomly distributed and fluctuate due to quantum mechanics, early versions of this scenario were called *chaotic inflation*.)

Here is another way to think of it (Linde 1994, 50). In GTR, expansion
is related to density. The higher the density of the universe, the greater
the expansion. Since energy is equivalent to mass ($E = mc^2$), the potential
energy of the scalar field increases expansion. Large inflaton field values
have more energy which in turn creates the exponential expansion
known as inflation. In a given universe, inflation ends when that potential energy reaches its (local) minimum. The lost energy is converted
into elementary particles.

Inflation itself is an unstable state of the inflaton field. As such, it will
tend to spontaneously decay (i.e., inflation stops). In some regions, the
energy field settles down to allow for a stable domain, like our observable
universe. Others with high inflaton values continue to inflate forever,
driving these other domains far away from our own. In the multiverse as a
whole, inflation never stops. It continues eternally in many domains but
decays in others. Inflation was therefore not merely a onetime event after
the Big Bang. Eternal inflation continues to produce bubble domains in a
grand fractal structure. Each domain, looking back in time, would see a Big
Bang that produced its own set of own values for the FTCs.

So how many universes are we talking about? Eternal inflation can be coupled with string theory to produce an estimate of 100^{500} (Smolin 2006, 158–162). In fact, this only accounts for universes with a positive cosmological constant. If we allow all values of Λ, says Smolin, this large but finite number becomes infinite.

2.4.3.3   *Disconnected Multiverse*

There are a few proposals for a multiverse in which the universes are and have always been causally and topologically disconnected from each other. Perhaps the best known is Tegmark's *Level IV*, a truly infinite multiverse (Tegmark 2003, 49). On this view, every mathematically consistent world actually exists. Each combination of all possible laws of nature, free parameters, and initial conditions is instantiated. No generating mechanism has been given for this. It is instead a speculative proposal for what physical reality might be like.[27]

2.4.4   *Multiverse: The Problems*

Enthusiasm for the multiverse seems to have peaked as a growing number of physicists see it as untestable. While I'll argue in Chapter 5 that this is not a fatal problem, there are several criticisms of the multiverse that do need to be considered. Some of these apply to each type of multiverse we've discussed; others depend on whether the number of universes is thought to be large but finite or truly infinite.

2.4.4.1   *Promissory Notes*

As we've seen, it's easy to move fine-tuning from one place to another, much harder to get rid of it entirely. It typically gets pushed under someone else's rug. This is what happened with early proposals for inflation[28] and now seems to be the case for more recent versions. Physicist Sean Carroll argues that eternal inflation requires the fine-tuning of entropy (2010, 334–338). For inflation to get up and running, the entropy of the preinflation universe would have to be extremely small compared to today (see Section 2.2.1). So while eternal inflation purports to explain the FTCs by generating a multiverse, it presupposes fine-tuning with respect to entropy—no small problem.

This is a common theme in the multiverse literature. There are no multiverse proposals that completely explain fine-tuning. All contain promissory notes for some FTCs, like entropy, or merely hope that fine-tuning will not arise again once a given proposal has been fleshed out.

2.4.4.2 An Infinite Multiverse

It is often assumed that if there are an infinite number of universes, then surely that explains fine-tuning. Unfortunately, when physicists use the word "infinite," they aren't always being precise. Eternal inflation is often said to produce an infinite number of domains, but given the details, it seems that this should be understood as finite and unbounded in the future—a potential infinite, like counting. At any given point in time, the number of universes will be finite. Nonetheless, let's take the claim at face value: there are an infinite number of universes presently instantiated in the multiverse.

Popular articles about fine-tuning don't talk about measure theory, but if we're going to compare infinite sets, it's unavoidable. Recall from our earlier discussion that any property with zero measure should almost certainly never occur, even though it is physically possible. Pencils should not balance on their tips for long periods of time. The oxygen in this room should not collect in the corner. To actually observe a property with measure zero (in a suitably realistic space of possibilities) would require a special explanation. In his discussion of observation selection effects and cosmological models, Earman puts it this way:

> But if the feature in question is unusual with a vengeance—measure zero—then the probability that it will be exhibited in some mini-world [i.e., a universe/domain] in the Ellis model is zero, and so no selection effect principle will suffice to explain away our puzzlement at encountering such a feature. (1987, 315)

As we saw earlier (Sections 2.3.3.3 and 2.3.4.2), it appears that the life-permitting ranges of the FTCs do have measure zero in an unbounded parameter space. Using measure as a guide to probability, this means that the probability of a universe like ours is zero. At best, zero probability multiplied by an infinite number of trials is undefined (Holder 2004, 148–149). An infinite number of universes do not guarantee that one like ours will be instantiated. Hence, the multiverse by itself does not explain the appearance of measure zero properties, such as the FTCs in a life-permitting universe.

2.4.4.3 A Finite Multiverse

We have only mentioned a couple of mechanisms for generating universes. There are dozens of ideas in the literature, most of which would produce a large but finite multiverse. One thing they lack is any sort of guarantee that the mechanism will produce enough life-friendly universes. Here's what I mean.

Earlier, I argued that we don't need *the* right measure in hand to know whether fine-tuning needs an explanation (Section 2.3.4.3). One doesn't need the precise odds for each case of fine-tuning to know that there is a problem to be solved. Here, the issue is different. Now, we're talking about whether a proposed explanation is acceptable. When we want to know what kinds of universes a generating mechanism produces, we now need a natural measure for that mechanism. This is analogous to saying that in order to know how often to expect 5s to be produced by a loaded die, we need a measure that captures the die's bias. In order to know whether a proposed universe-generating mechanism *solves* the problem of fine-tuning, we need a measure for that mechanism.

So what kind of measure would one need? First, it would have to allow for a variety of values for the FTCs. As Collins argues, this is far from guaranteed:

> [There] must be some mechanism that allows for enough variation in the parameters of physics to account for the fine-tuning. This would require that the fundamental structure of physical law have the right form, as might be the case in string theory but is not the case, for example, in the typical grand unified theories that are being studied today. ... As Joseph Polchinski notes ... there is no reason to expect a generic field to have an enormous number of stable local minima of energy, which would be required if there is to be a large number of variations in the parameters of physics among universes. (Collins 2002, 7)

Second, if the multiverse is biased against life-permitting universes, then the fact that there are lots of universes out there will not explain fine-tuning. Howsoever universes are produced, those with life-permitting FTCs must be of positive measure, which is to say there is a nonnegligible probability that such universes will exist in the multiverse. In short, all of the multiverse proposals are incomplete without a natural measure showing the distribution of universes produced.

2.4.4.4   *Boltzmann Brains*

The most intriguing issue for all versions of the multiverse has to do with Boltzmann brains. The problem goes back to the nineteenth-century work of Ludwig Boltzmann on statistical mechanics (Carroll 2010, 221–224). In

Boltzmann's early view, the universe should for the most part be close to equilibrium. High-entropy states are far more likely than low-entropy ones, so the universe remains close to equilibrium most of the time. On occasion, random fluctuations occur in which order spontaneously increases (entropy decreases). By the second law of thermodynamics, this orderly state will then decay back into equilibrium. The reason we find our universe in an orderly state, he thought, is simply a matter of chance. We reside in one of those random fluctuations with low entropy, working its way back to equilibrium.

The problem with this story is that our universe has *very* low entropy compared to what it could be, much less than we need. By analogy, say that each of my students is given a bucket of multicolored sand. They throw all the sand up in the air, see how it lands on the floor, and then repeat the process. The vast majority of the time, the sand falling to the ground will be randomly distributed. No discernible pattern or order will be evident. On occasion though, the sand will produce a pattern, something like a letter E or Roman numeral, for example. If we allow this process to go on indefinitely, then once in a very great while more interesting structures will emerge. As the number of throws gets extremely large, eventually one will produce a version of the Mona Lisa. We know this will happen somewhere down the line because that's just how random processes and large numbers work. Eventually, all possible combinations of sand will be produced. On average, though, there will be far more basic shapes produced than portraits—low-order outcomes as compared to high order.

Back to cosmology. If the order in the universe is merely the product of random fluctuations in the arrangement of matter, then it is far more likely that a galaxy would form out of the chaotic void than an entire universe of galaxies. More likely still that one solar system would form rather than a low-entropy galaxy since modest entropy fluctuations are more common than low-entropy ones. In fact, to account for your personal experience, all that needs to exist is a functioning brain. What seems like the experience of an embodied person living in a large ecosystem could instead be the hallucination of a disembodied brain floating in the void. All of your "memories" and present experience could be illusions which will end shortly. As Descartes showed, this is a very difficult claim to disprove. Not only is this scenario physically possible, it is overwhelmingly probable if Boltzmann's view is correct. Since a single brain contains less entropy than an entire planet like ours, random fluctuations from equilibrium have produced far more hallucinating, disembodied brains than habitable universes. Odds are that you are one of them. Sorry.

The Boltzmann brain story is a *reductio ad absurdum*. If one's physical theory indicates that the best explanation for my own subjective experience, including memories, is that I am a disembodied brain temporarily hallucinating in the void (rather than a real person currently sitting at my desk), that's a problem for one's theory. A set of beliefs known to be grounded on an illusion contains its own defeater. Any theory that leads to radical skepticism about one's experience would invalidate whatever evidence one had for the theory itself. In other words, once you believe it, you probably shouldn't.

If all that seems farfetched, there is a less eccentric way of looking at it. On the order-is-due-to-random-fluctuations view, low-order fluctuations are far more common than those with high order and low entropy. Most of the former will be tiny islands of order within the vast equilibrium void—a bit of random order over here and a bit over there. Since those universes are so much more common, that's what we would expect our universe to look like: a small habitable bubble with nothing else around. But that is not what we now observe. We are not in an isolated solar system or galaxy. The entire observable universe has extremely low entropy, an observation that counts against Boltzmann's view.

As you might have guessed, the Boltzmann brain is not merely a relic of the nineteenth century.[29] Expanding universes with a positive cosmological constant—like our own—will in the very distant future end up in a high-entropy equilibrium state. Instead of free-floating atoms as Boltzmann would have thought, there will be nothing but elementary particles and radiation, all of which are subject to thermal and quantum fluctuations. Most of these fluctuations will be so minor that an observer wouldn't notice. But since each such universe will, so far as we know, continue in this state forever rather than recollapsing in a Big Crunch, there will be no time at which these fluctuations will end. As the aeons pass, even macroscopic objects will on very rare occasions randomly emerge, including the whole solar systems and galaxies. Of course, as Boltzmann knew, a solar system contains a lot of order. Fluctuations that are close to equilibrium will occur far more often than low-entropy ones, like a life-permitting planet. But as we saw before, hallucinating brains contain less order than a planet like ours. There will therefore be far more Boltzmann brains produced in the long run than "standard observers" like you and me sitting comfortably in a stable galaxy.

So doesn't this count against the view that we live in an expanding universe with a positive cosmological constant? If those universes produce a lot more Boltzmann brains than standard observers, we seem to have landed in the same *reductio* as before. As Page has argued, the answer is no, not if we live in a universe rather than a multiverse. The proliferation of Boltzmann brains is restricted in a universe like ours,[30] but not in a multiverse produced by eternal inflation, which is the most commonly accepted mechanism for producing a multiverse (2008b). While our universe might one day produce some Boltzmann brains, they will not vastly outnumber standard observers. In other words, in a universe like this one, you probably aren't a Boltzmann brain. (Cheer up.) In a multiverse on the other hand, there should be far more Boltzmann brains that currently exist than standard observers. Hence, the Boltzmann brain *reductio* seems to be a problem for a multiverse but not a universe. Something must be wrong.

What is wrong is that even a finite multiverse is too big. Multiverse advocates agree that if nature had only one chance to get all the FTCs to line up just so, a habitable universe is wildly improbable. One can expand the probabilistic resources of nature by positing a multiverse rather than a universe—more rolls of the dice, as it were—but in solving one problem, others are created, such as Boltzmann brains. The multiverse is therefore not the easy remedy for fine-tuning as is so commonly assumed.

2.4.4.5   Conclusions

Let's take care of some loose ends. First, as Ernan McMullin has pointed out, there is an MSP below the surface that we have not encountered: the principle of indifference (2008). It says that our circumstances should be generic rather than special from a naturalistic point of view. Nature should be indifferent to our existence. For example, Earth should be one of many planets, including those capable of supporting life. Our sun should be one of many stable, slow burning stars, and so on. Older cosmologies with Earth at the center of the universe are the antithesis of this principle. McMullin argues that this principle drives fine-tuning research at least in part. The multiverse "illustrates the speculative—some would say extreme—lengths to which cosmologists will go in their efforts to maintain the indifference principle" (McMullin 1993, 386). This is another illustration of the top-down pressure that MSPs exert on theory choice and research that we discussed in Chapter 1.

Second, some have questioned whether the multiverse is a scientific proposal. Increasingly, physicists say no; it's not a testable hypothesis and therefore cannot be science. John Polkinghorne calls it a metaphysical posit. While McMullin was not a multiverse advocate, he did believe that it is scientific, a case of inferring a cause from its effect or *retrodiction* (2008, 89). In Chapter 5, I'll argue for a different view. The real question is not whether the multiverse and/or design count as science. Demarcation arguments unsuccessfully try to force precision onto an intrinsically fuzzy issue. The right question is whether they count as *good* science. Both the multiverse and theistic design are explanatory, and that's good. Neither is directly testable, and that's bad. Hence, if a testable explanation for fine-tuning were to come along, it would tend to trump both design and the multiverse.

Third, nothing in this chapter entails that theism and the multiverse are logically contradictory notions. A small number of theists have argued in favor of the multiverse, including Don Page (2008a). While I don't find the arguments for a theistic multiverse persuasive, I agree that it is a logically consistent idea.

Everything in this chapter pertains to the best theories currently in play. Future science might find a way to explain fine-tuning without the problems we've discussed, but as for now, there is not even a proposal that adequately explains fine-tuning in naturalistic terms. Given the parallels between the fine-tuning of artifacts (like a car or computer) or artificial environments (like a saltwater aquarium) and the fine-tuning of the universe, theists are within their epistemic rights to conclude that design is the best explanation at present. This is not a "proof" for the existence of God. I don't think there are any of those. Nonetheless, a case can be made that theistic design is a better explanation than its naturalistic rivals.

Notes

1   That's the G in the equation $F = G\dfrac{m_1 m_2}{r^2}$. While the force of gravity depends on mass and distance, the gravitational constant does not. G is the same everywhere in the universe, hence the name 'universal gravitation.'

2   A related set of examples deals with the precise laws of nature needed for life. Although not strictly a case of fine-tuning, Robin Collins argues that if any one of the five laws were eliminated, then complex, self-reproducing life would not be possible: "(1) a universal attractive force, such as gravity; (2) a force relevantly similar to that of the strong nuclear force, which binds protons and neutrons together in the nucleus; (3) a force relevantly similar to that of the electromagnetic force; (4) Bohr's Quantization Rule or something similar; (5) the Pauli Exclusion Principle" (2009, 211). All but the first are essential for chemistry. Stable atoms and complex molecules would not be possible in a classical world with Newtonian forces. Polkinghorne adds the need for gravity to be an inverse-square law. If it were inverse cubed, planets could not have closed, stable orbits, at least in solar systems with multiple planets. Perturbations would either send each planet spiraling into the star or off into space.

3 "In general, it makes hardly any difference what is assumed as the initial state of the universe, since all subsequent changes must take place in accordance with the laws of nature. It is difficult to imagine any initial state from which the same effects could not be deduced by the same laws of nature, though perhaps with more effort. Thanks to these laws of nature, matter successively takes on all the forms of which it is capable. Consequently, if we consider these forms in proper order, we shall finally be able to arrive at the form of the world we currently live in. This is so true that there can be no fear of error as the result of making a false assumption about the initial conditions" (*Principles of Philosophy*, III, 47).

4 Technically, what follows describes a *phase space*. If the universe only contained one particle, the state space required to capture all of these possibilities would have six dimensions: three to capture the position of the particle with spatial coordinates $\langle x, y, z \rangle$ and three to record its momentum along each axis $\langle p_x, p_y, p_z \rangle$. For each additional particle, we will need six more dimensions in order to capture all of this information.

5 Penrose has suggested a new constraint on spacetime singularities like the Big Bang that would impose initially low entropy. Critics, like Larry Sklar, argue that Penrose's calculation leaves out the effects of gravity (Sklar 2009). That's fine when dealing with atomic interactions, but not when considering the early universe and entropy.

6 More precisely, the parameter at issue here is the *effective dark energy density*, which is only part of the cosmological constant. Moreover, Collins uses a very conservative estimate in his paper. The more typical value given by physicists is 10^{123} (Collins, private correspondence).

7 Robin Collins discusses two other examples of "seriously problematic" cases of fine-tuning (2003, 191–195): the strong nuclear force and gravity vis-à-vis the creation of stable stars.

8 Such as Timothy O'Connor (2008) or Michael Rea (2004).

9 Rodney Holder suggests another approach to the coincidence objection (private correspondence). This universe is special insofar as it contains objective value. A universe with life, some measure of happiness, art, knowledge, and virtue has real value to it, beyond our subjective feelings about such things. Other possible universes are therefore not on a par with this one. A universe with value requires explanation that others do not.

10 See Weisberg (2005), Nunley (2010), and White (2011).

11 See Earman (1992), Monton (2006), and Kotzen (2012). The problem of old evidence is that observations made years ago can help confirm a new hypothesis. Recall that if $Pr(e|h_1) > Pr(e|h_2)$, then e favors h_1 over h_2. Let e_M = 'the perihelion of Mercury shifts around the Sun 5600 seconds of arc per century,' h_N = Newtonian physics and h_{GTR} = GTR. This is a famous example since according to Newton's laws, the precession of Mercury's perihelion should be

slower by 43 seconds of arc per century. The fact that GTR gets it right was considered an important confirmation of GTR. Notice, however, that long before Einstein proposed GTR, e_M was known to be true, that is, it was part of background knowledge k. Unfortunately, this entails that $Pr(e_M|h_N \& k) = Pr(e_M| h_{GTR} \& k) = 1$, and so e_M *cannot* count as evidence in favor of GTR under the law of likelihoods, although it clearly should count as evidence. That's the problem. If known observations are included in k, then old evidence cannot favor a new theory over an older rival. There are several proposals for getting around this problem, none of which has gained a clear following. Strictly speaking, Sober is not dealing with old evidence as it is normally defined since 'you exist' does not entail 'you observe that the constants are right.' Nonetheless, the debate over whether one's existence should be a given is in the same neighborhood as questions about old evidence and background knowledge. It is very hard to resolve such matters without begging the question.

12   If one isn't looking for a way out, then I suggest Kotzen (2012) as the best solution to date. Kotzen argues that one's own observation of fine-tuning rightly belongs with the evidence rather than with the background knowledge, when comparing the probabilities. Others, like Rodney Holder, argue that if our existence is what needs to be explained, then it cannot be treated as background knowledge (private correspondence).

13   I'm assuming the probability distribution is spread uniformly with the measure. That's a simplification. The measure has to do with relative sizes of the space. The probability distribution need not match. Think of an irregular, asymmetric die with different size faces. The measure would capture the relative size of the faces. But let's also say the die is loaded to favor one of the smaller sides. Even though the measure would be small for that side, the probability distribution would be large because of the bias in the die. For our purposes, I'll use the term "probability measure," combining both measure (size) and distribution.

14   For examples, see Holder (2001, 346).

15   No matter how curved space was immediately after the Big Bang, inflation stretches it to the point of being very close to Euclidean (Penrose 2004, 747). By analogy, an extremely large balloon will have an irregular surface before it is inflated. A flea walking on it would have to negotiate all sorts of wrinkles and valleys. After the balloon is inflated, the surface will appear locally flat to the flea. The initial irregularities are smoothed away.

16   "In this way a uniform probability distribution in the canonical measure would explain the flatness problem of cosmology…" (Hawking and Page 1988, 803–804). Would, that is, if Gibbons *et al.* had been able to show that *inflation* accounted for flatness. Hawking and Page proved instead that although flatness is held in almost all of the models, it was not necessarily due to inflation.

17   Such arguments are common in nonlinear dynamics. Sauer *et al.* (1991) embedding theorems are a particularly important example.

18   See Koperski (2005) for more.

19   Although there are variations, the most famous is that every number greater than 2 is the sum of 3 primes (http://mathworld.wolfram.com/GoldbachConjecture.html).

20   At the beginning of this section, I mentioned other interpretations of probability than an objective one. Rodney Holder points out that if we were to use epistemic probabilities instead, the opinions of experts would be an important part of one's basis for belief and hence play a direct role in the probability calculations (private correspondence).

21   Not everyone believes that inflation will turn out to be a good explanation. "Undoubtedly the inflationary paradigm has been very fashionable for almost two decades now. But fashionability does not equate with epistemic justification, and it is certainly not a guarantor of truth" (Earman and Mosterin 1999, 2–3). Paul Shellard lists over one hundred versions of inflation (2003, 764), none of which are as elegant as the initial proposal seemed to be.

22   As cosmologist Robert Brandenberger, the cosmological constant (or something that plays essentially the same role) will have to be tuned precisely to allow a graceful exit from the inflationary epoch: "The field which drives inflation … is expected to generate an unacceptably large cosmological constant *which must be tuned to zero by hand*. This is a problem which plagues *all* inflationary universe models" (Craig 2003, 172). See Holder (2004, 135–137) for a nice description of the whac-a-mole problem in the history of inflation.

23   Philosophers generally agree—well, me and Ryan Nichols anyway—that *Star Trek* often presents old philosophical questions in new and interesting ways. It also seems to consistently get the answers to those questions wrong. My personal favorite is from the *Voyager* series when a new individual, Tuvix, is created in a transporter accident—it's always a transporter accident—by melding two other characters, Tuvok and Neelix. The two are transported, but only Tuvix emerges. After several weeks, Capt. Janeway orders Tuvix to undergo separation against his will, a clear violation of his right to life.

24   This bucks the majority view regarding black holes and entropy. Most physicists believe that in the process from black hole formation to final evaporation, the total entropy *increases* (Carroll 2010, 303).

25   "[We] must continue to bear in mind that these proposals rely for the most part on unknown physics together with extrapolations of presently known physics to realms far beyond where its reliability is assured" (Ellis *et al.* 2004).

26   The energy of the inflaton field generates gravity; the antigravitational effect is stronger by a factor of three (Davies 2006, 59).

27   A close cousin of this idea is David Lewis's *modal realism* (2001), although Lewis's view is motivated by philosophical issues rather than physics.

28   "Unfortunately the necessary slow-rollover transition requires the fine tuning of parameters (notably the energy-density, the quantity which occasioned the

apparent need for fine-tuning in the original Big Bang model); calculations yield reasonable predictions only if the parameters are assigned values in a narrow range. Most theorists (including both of us) regard such fine tuning as implausible. The consequences of the scenario are so successful, however, that we are encouraged to go on in the hope that we may discover realistic versions of grand unified theories in which such a slow-rollover transition occurs without fine tuning" (Steinhardt 1984, 127).

29   For a related argument using "fluctuation observers" like Boltzmann brains, see Collins (2009, 266–267, 2012).

30   This is because, Page calculates, our universe will most likely not be able to support observers of any kind after 20 billion years. Hence, there will be no vast epoch of time in which Boltzmann brains come to dominate this universe.

References

Barr, Stephen M. 2010. "The Multiverse and the State of Fundamental Physics Today." In *Science and Religion in Dialogue*, edited by Melville Y. Stewart, 2:911–927. Malden: Wiley-Blackwell.

Barrow, John D. 2002. *The Constants of Nature*. New York: Pantheon Books.

Barrow, John D., and Frank Tipler. 1986. *The Anthropic Cosmological Principle*. New York: Oxford University Press.

Callender, Craig. 2004. "Measures, Explanations and the Past: Should 'Special' Initial Conditions Be Explained." *British Journal for the Philosophy of Science* 55 (2): 195–217.

Carroll, Sean. 2010. *From Eternity to Here: The Quest for the Ultimate Theory of Time*. New York: Dutton.

Collins, Robin. 2002. "The Argument from Design and the Many-Worlds Hypothesis." In *Philosophy of Religion: A Reader and Guide*, edited by William Lane Craig, 130–148. New Brunswick: Rutgers University Press.

Collins, Robin. 2003. "Evidence for Fine-Tuning." In *God and Design: The Teleological Argument and Modern Science*, edited by Neil A. Manson, 178–199. New York: Routledge.

Collins, Robin. 2005. "Hume, Fine-Tuning, and the 'Who Designed God?' Objection." In *In Defense of Natural Theology: A Post-Humean Assessment*, edited by James Sennett and Douglas Groothius, 175–199. Downers Grove: InterVarsity Press.

Collins, Robin. 2009. "The Teleological Argument: An Exploration of the Fine-Tuning of the Universe." In *The Blackwell Companion to Natural Theology*, edited by William Lane Craig and J.P. Moreland, 202–281. Chichester: Wiley-Blackwell.

Collins, Robin. 2012. "Modern Cosmology and Anthropic Fine-Tuning: Three Approaches." In *Georges Lemaître: Life, Science and Legacy*, edited by Rodney D. Holder and Simon Mitton. Berlin/Heidelberg: Springer.

Craig, William Lane. 2003. "Design and the Anthropic Fine-Tuning of the Universe." In *God and Design: The Teleological Argument and Modern Science*, edited by Neil A. Manson, 155–177. New York: Routledge.

Craig, William Lane, and James Sinclair. 2009. "The Kalam Cosmological Argument." In *The Blackwell Companion to Natural Theology*, edited by William Lane Craig and J.P. Moreland, 101–201. Chichester: Wiley-Blackwell.

Davies, Paul. 1982. *The Accidental Universe*. Cambridge: Cambridge University Press.

Davies, Paul. 2006. *The Goldilocks Enigma: Why Is the Universe Just Right for Life?* Boston: Houghton Mifflin.

Earman, John. 1987. "The SAP Also Rises: A Critical Examination of the Anthropic Principle." *American Philosophical Quarterly* 24 (4): 307–317.

Earman, John. 1992. *Bayes or Bust?: A Critical Examination of Bayesian Confirmation Theory*. Cambridge: MIT Press.

Earman, John, and Jesus Mosterin. 1999. "A Critical Look at Inflationary Cosmology." *Philosophy of Science* 66 (1): 1–49.

Ellis, G.F.R., U. Kirchner, and W.R. Stoeger. 2004. "Multiverses and Physical Cosmology." *Monthly Notices of the Royal Astronomical Society* 347 (3): 921–936.

Evrard, G., and P. Coles. 1995. "Getting the Measure of the Flatness Problem." *Classical and Quantum Gravity* 12: 193–197.

Gibbons, G.W., S.W. Hawking, and J.M. Stewart. 1987. "A Natural Measure on the Set of All Universes." *Nuclear Physics B* 281 (3–4): 736–751.

Halvorson, Hans, and Helge Kragh. 2013. "Theism and Physical Cosmology." In *The Routledge Companion to Theism*, edited by Stewart Goetz, Victoria Harrison, and Charles Taliaferro, 241–255. New York: Routledge.

Hawking, Stephen W., and Don N. Page. 1988. "How Probable Is Inflation." *Nuclear Physics B* 298: 789–809.

Heaviside, Oliver. 1894. *Electromagnetic Theory*. Vol. 2. 3 vols. "The Electrician" Series. London: "The Electrician" Printing and Publishing Company Limited. http://catalog.hathitrust.org/Record/002241259. Accessed December 3, 2012.

Holder, Rodney D. 2001. "The Realization of Infinitely Many Universes in Cosmology." *Religious Studies* 37 (3): 343–350.

Holder, Rodney D. 2004. *God, the Multiverse, and Everything: Modern Cosmology and the Argument from Design*. Aldershot: Ashgate.

Kirchner, U., and G.F.R. Ellis. 2003. "A Probability Measure for FLRW Models." *Classical and Quantum Gravity* 20: 1199–1213.

Koperski, Jeffrey. 2005. "Should We Care about Fine-Tuning?" *British Journal for the Philosophy of Science* 56 (2): 303–319.

Kotzen, Matthew. 2012. "Selection Biases in Likelihood Arguments." *British Journal for the Philosophy of Science* 63 (4): 825–839.

Lewis, David K. 2001. *On the Plurality of Worlds*. Malden: Blackwell Publishers.

Linde, Andrei. 1994. "The Self-Reproducing Inflationary Universe." *Scientific American* 271 (5): 48.

Manson, Neil A. 2000. "There Is No Adequate Definition of 'Fine-Tuned' for Life." *Inquiry* 43: 341–352.

McGrew, Timothy, Lydia McGrew, and Eric Vestrup. 2001. "Probabilities and the Fine-Tuning Argument: A Sceptical View." *Mind* 110 (440): 1027–1038.

McMullin, Ernan. 1993. "Indifference Principle and Anthropic Principle in Cosmology." *Studies in the History and Philosophy of Science* 24 (3): 359–389.

McMullin, Ernan. 2008. "Tuning Fine-Tuning." In *Fitness of the Cosmos for Life: Biochemistry and Fine-Tuning*, edited by John D. Barrow, 70–94. Cambridge: Cambridge University Press.

Monton, Bradley. 2006. "God, Fine-Tuning, and the Problem of Old Evidence." *British Journal for the Philosophy of Science* 57: 405–424.

Nunley, Troy. 2010. "Fishnets, Firing Squads, and Fine-Tuning (Again)." *Philosophia Christi* 12 (1): 160–178.

Oberhummer, H.H., A. Csótó, and H. Schlattl. 2000. "Fine-Tuning of Carbon Based Life in the Universe by Triple-Alpha Process in Red Giants." *Science* 289: 88–90.

O'Connor, Timothy. 2008. *Theism and Ultimate Explanation: The Necessary Shape of Contingency*. Malden: Blackwell.

Page, Don. 2008a. "Does God So Love the Multiverse?" *Eprint arxiv.org/abs/0801.0246v5*. http://arxiv.org/abs/0801.0246v5. Accessed January 9, 2013.

Page, Don. 2008b. "Is Our Universe Likely to Decay within 20 Billion Years?" *Physical Review D* 78 (6). DOI:10.1103/PhysRevD.78.063535.

Penrose, Roger. 1989. *The Emperor's New Mind*. Oxford: Oxford University Press.

Penrose, Roger. 2004. *The Road to Reality*. New York: Knopf.

Penrose, Roger. 2006. "Before the Big Bang: An Outrageous New Perspective and Its Implications for Particle Physics." *Proceedings of European Particle Accelerator Conference (EPAC 06)*. Edinburgh, Scotland, pp. 2759–2767.

Rea, Michael C. 2004. *World Without Design: The Ontological Consequences of Naturalism*. New York: Oxford University Press.

Sauer, Tim, James A. Yorke, and Martin Casdagli. 1991. "Embedology." *The Journal of Statistical Physics* 65: 579–616.

Shellard, Paul. 2003. "The Future of Cosmology: Observational and Computational Prospects." In *The Future of Theoretical Physics and Cosmology: Celebrating Stephen Hawking's Contributions to Physics*, edited by S.W. Hawking, G.W. Gibbons, E.P.S. Shellard, and S.J. Rankin, 755–777. New York: Cambridge University Press.

Sklar, Lawrence. 1993. *Physics and Chance: Philosophical Issues in the Foundations of Statistical Mechanics*. Cambridge/New York: Cambridge University Press.

Sklar, Lawrence. 2009. "Philosophy of Statistical Mechanics." In *Stanford Encyclopedia of Philosophy*, edited by Edward N. Zalta. Stanford: Metaphysics Research Lab, Center for the Study of Language and Information, Stanford University.

Smolin, Lee. 1999. *The Life of the Cosmos*. New York: Oxford University Press.

Smolin, Lee. 2006. *The Trouble with Physics*. Boston: Houghton Mifflin.

Smolin, Lee. 2007. "Scientific Alternatives to the Anthropic Principle." In *Universe or Multiverse?*, edited by Bernard Carr, 323–366. Cambridge: Cambridge University Press.

Sober, Elliot. 2003. "The Design Argument." In *God and Design: The Teleological Argument and Modern Science*, edited by Neil A. Manson, 25–54. New York: Routledge.

Sober, Elliot. 2009. "Absence of Evidence and Evidence of Absence: Evidential Transitivity in Connection with Fossils, Fishing, Fine-Tuning, and Firing Squads." *Philosophical Studies* 143 (1): 63–90.

Steinhardt, Paul J. 1984. "The Current State of the Inflationary Universe." *Comments on Nuclear and Particle Physics* 12: 273–286.

Steinhardt, Paul J. 2011. "The Inflation Debate." *Scientific American* 304 (4): 36–43.

Swinburne, Richard. 1979. *The Existence of God*. Oxford: Oxford University Press.

Tegmark, Max. 2003. "Parallel Universes." *Scientific American* 288: 40–51.

Weisberg, J. 2005. "Firing Squads and Fine Tuning—Sober on the Design Argument." *British Journal for the Philosophy of Science* 56: 809–821.

White, R. 2011. "What Fine-Tuning's Got to Do with It: A Reply to Weisberg." *Analysis* 71 (4): 676–679.

Wilson, Mark. 2010. "What Can Contemporary Philosophy Learn from Our 'Scientific Philosophy' Heritage." *Noûs* 44 (3): 545–570.

3

Relativity, Time, and Free Will

3.1 Physics and Freedom

If you have ever been to an amusement park, you have probably seen the unfortunate kid-on-a-leash. In order to keep a rambunctious child from wandering away, parents put him—it's always a "him"—in a harness attached to a nylon cord or leash. It's usually not called a "leash," of course, since leashes are for pets. The idea is to allow the child to have some freedom, but not too much. (The child's own sense of freedom is somewhat limited, one would think.)

On a more theoretical plane, physics has often undermined our notion of free will, and philosophers have been concerned about this for many reasons. One is that freedom seems to be a necessary condition for moral responsibility. If two very strong men grab my hand and force me to punch you in the nose, I am not guilty of assault; I had no choice in the matter. In terms of religion, acts of piety are only praiseworthy if one has the ability to do otherwise. Unless you have the ability to choose, God has no more reason to be pleased with your giving to the poor than he has reason to be pleased with your having DNA. For this and other reasons, free will matters to theists.

As I mentioned, physics has not always been helpful in this regard. Pierre Laplace famously held that given the position and momentum of every atom and enough computational capacity, one could use Newton's laws to predict the exact state of the universe anytime in the future.[1] If the behavior of all things, including the atoms in our own bodies, is wholly determined by the laws of nature, then there doesn't appear to be any room left for free will.

The Physics of Theism: God, Physics, and the Philosophy of Science,
First Edition. Jeffrey Koperski.

No one worries about Laplacian determinism anymore since Newtonian physics was replaced by quantum mechanics.[2] But the story does not end there. The other pillar of 20th-century physics, Einstein's theory of relativity, poses another challenge to freedom and to our understanding of time itself. Most of us assume that there is a fundamental difference between the past, present, and future. What has happened is fixed and cannot be changed. The future, on the other hand, is open and allows for a vast range of possibilities. We exist in the present, the moving demarcation between past and future. Unless a clever philosopher has told you otherwise, this is most likely your view. Many physicists and philosophers believe that the special and general theories of relativity (STR and GTR, respectively) have shown this view of time to be false. Time, they tell us, does not flow. There is no ontological difference between past and future, and there is no special point on the timeline that is "the present." From the physicist's point of view, almost all of our commonsense beliefs about time are based on an illusion. Odds are no one told you *that* in freshman physics.

When I talk about free will in this chapter, I have a robust type of freedom in mind. The key idea is that future events are in part determined by our choices and those choices are not themselves determined by anything in the past. For example, my mp3 player recently broke. (Okay, I dropped it.) I am currently looking into new players from four different companies. I believe that, at this instant, there is no fact of the matter about which of the four I will buy.[3] The decision is up to me and I haven't yet decided. Which of these possibilities comes to pass will be determined by my free decision. No matter what the biological, psychological, physical, and metaphysical facts are prior to this decision, my choice of mp3 player is ultimately up to me. This view is what is technically known as *libertarian* freedom.[4] If I choose one way, the future will develop accordingly, but I have it in my power to choose something else thus bringing about a slightly different future.

Theists have a stake in all this for several reasons. If there is no real freedom of will, then the idea that one has any sort of personal responsibility for one's choices is undermined. Mother Teresa's actions are not praiseworthy and Hitler's are not blameworthy if they could not have done otherwise. The metaphysics of time is also more important than it might seem. John Lucas argues that if time is static and God is timeless, then reality consists of two unchanging entities in a fixed relation to one another (2008, 279). Prayer makes little sense in such a scenario. God cannot hear a prayer *and then* alter some event in the future if both God and the future are fixed.

The timeline of the world would simply exist without change, much as the Greek philosopher Parmenides once taught.

One might be tempted to simply reject all this out of hand. "Einstein didn't believe in time? Oh well, geniuses can be a little eccentric." Before going that route, consider how many other commonsense beliefs have been overthrown by physics. Start with the view that the earth is the center of the universe. The sun revolving around the earth is perfectly empirical. If you had not been taught otherwise, you would almost certainly be a geocentrist. Would anyone believe that we live at the bottom of an ocean of air pressing on our bodies at 14.7 pounds per square inch without the advent of modern science? Photons, black holes, X-rays, etc. What once seemed absurd is now common knowledge. Let's now consider whether our views about time and freedom also need to be changed in light of current physics. Section 3.2 will explain how STR leads to the idea that there is no flow of time. Section 3.3 analyzes several proposals that reintroduce a classical view of time without violating relativity. In Section 3.4, I suggest two ways in which the philosophy of science can add some helpful perspective to the debate.

3.2   STR and the Nature of Time

3.2.1   *The Metaphysics of Time*

Time seems to flow from the past through the present and into the future. Science fiction aside, we tend to believe that the present is fully real but not the future or past. There is no sense in which the 13-year-old Jeff Koperski is sitting in the stupefying boredom of Mrs. Elsner's English class. Those events are memories, but they no longer literally exist. The past has no ontological weight. It's not "out there" somewhere.

The future doesn't exist either in that there are many ways the future might yet go. I'll be raking a lot of leaves this fall, unless a big wind blows them down the street. (One can only hope.) Which of the possibilities comes to pass is contingent on events and choices that have not yet happened. There is no sense in which a future Jeff Koperski exists and is cursing the leaves in his yard. On the commonsense view, time travel isn't a technological problem; it's an ontological one. We can't build a time machine because there aren't any other times in existence to which one could go. It's kind of like saying that we can't reach the end of a rainbow. It's not forbidden; there just isn't any place that is the-end-of-the-rainbow.

The notion that there is a metaphysical difference between past and future is known as the *dynamic* or *A-series* view of time. Although there are different types of A-series, they all include absolute becoming (i.e., events only fully come into existence when they reach the present) and the view that time has an objective flow to it. The most common form of A-series, known as *presentism*, adds that only the present exists.[5] The main rival to presentism is *eternalism*, also known as the *B-series*, *block universe*, or *static* view of time: each moment fully exists without reference to past or future. So the event denoted by 'Jeff Koperski graduates with a Ph.D. on June 7, 1997' is just as real in the B-series as 'Jeff Koperski types a draft of Chapter 3 on June 7, 2013.' There is no past or future in the block universe. There is only before and after with respect to a given point on the timeline. Physicist James Jeans describes it this way:

> In this case our consciousness is like that of a fly caught in a dusting-mop which is being drawn over the surface of the picture; the whole picture is there, but the fly can only experience the one instant of time with which it is in immediate contact, although it may remember a bit of the picture just behind it, and may even delude itself into imaging it is helping to paint those parts of the picture which lie in front of it. (Lockwood 2005, 54–55)

In fact, Jeans's characterization of B-series time isn't quite right, as philosopher Bradley Monton points out (private correspondence). The notion of being moved over the surface of the picture implies an objective passage of time, which is incompatible with the B-series view. There is no moving present, no one place on the picture where the fly can be found. Einstein himself was more succinct: "[The] distinction between past, present, and future is only a stubbornly persistent illusion" (Speziali 1979, 51).

With the relevant categories defined, let's consider why STR seems to support a static rather than dynamic view of time.

3.2.2   Relativity and the Present

Einstein's development of special relativity was based on two fundamental ideas.[6] (1) From classical mechanics, the laws of nature work the same in every inertial frame (i.e., motion at a constant velocity without acceleration). Hence, the way ping-pong balls behave in your garage (reference frame A) is the same way they behave in a bullet train with a constant velocity (reference frame B). (2) From electrodynamics, the speed of light is a constant c

which is the same for all observers, regardless of whether one is approaching the source of light or retreating from it.

Neither sounds terribly radical. How could these innocent looking postulates change the metaphysics of time?

First, we must realize that not all velocities add the way they do in Newtonian mechanics. Say that Marcus is on a flatbed train car moving at a constant speed of 90 km/hour and that Andrew is standing on the ground alongside the tracks watching the train go by. Since Marcus is traveling at a constant velocity, the moving train counts as one reference frame. Andrew is in a second reference frame. Say that Marcus throws a baseball toward the front of the train just as he's passing Andrew (Figure 3.1).

From Marcus's point of view, the ball travels at 75 km/hour. But from Andrew's point of view, the ball moves much faster: 165 km/hour, the added speed of the ball and train. So far, so good. But what if instead of a ball, Marcus aims a laser pointer toward the front of the train? How will the relative motion affect the speed of light? In this case, the velocities do not add, according to STR. Marcus will measure the speed of light moving away from him to be $c = 3 \times 10^8$ m/second and so will Andrew. It doesn't matter how fast one observer is moving relative to another, both will detect the speed of light to be precisely c.

Now say that Marcus is inside an enclosed train car moving at a constant speed of 90 km/hour (Figure 3.2). This particular car is unusual in that its walls are made of one-way mirrors. People outside the train can see inside, but Marcus can only see his reflection. Andrew is again standing on the

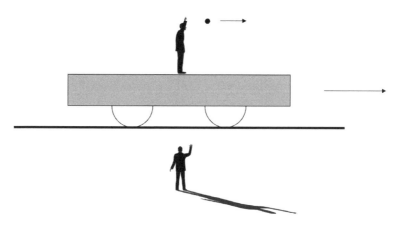

Figure 3.1   Velocities add.

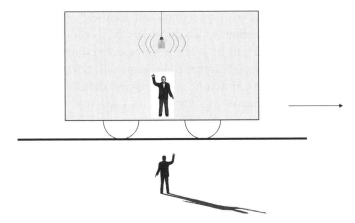

Figure 3.2 STR flashbulb experiment.

ground alongside the tracks. Now say that Marcus is in the very center of the car along with a flashbulb. Finally, Marcus coincidentally sets off the flashbulb when he crosses the precise spot where Andrew is standing. The light from the flashbulb moves at 3×10^8 m/second from the center of the car toward both the front and the back. Call the event of the light hitting the front of the train car L_f and the light hitting the back L_b. From Andrew's point of view, the light travels equally fast in both directions, but the front of the car is moving away from the wavefront while the back of the car is moving toward it. Hence, L_b happens slightly before L_f. From Marcus's point of view, the light travels equal distances in both directions. According to principle (i), regardless of whether the train is sitting in the station or moving at the maximum speed, Marcus's observations inside the car must be the same in both cases. If the light is moving at c in both directions and it has an equal distance to cross in both directions, Marcus will observe that the light hits both walls at the same time. L_b is simultaneous to L_f.

Marcus says the two events are simultaneous and Andrew says not. What every freshman physics student wants to know is, which one is right? The answer according to STR is that neither one has *the* correct answer. There is no one right answer; all inertial reference frames are on a par. Without a privileged frame, there is no place to stand in which one could know the truth about simultaneous events. There is no truth about simultaneous events, only different observations from different frames. If you're tempted to say that Andrew must be right since he's at rest, consider that Andrew is hurtling through space on a very large spheroid. It would be easy to recast this thought experiment with spaceships and glass walls where the reader would have no intuition about which ship is "really" moving. In STR, no one is really at rest. One can only be at rest with respect to some object or reference frame. What counts as simultaneous events differs from one reference frame to another. More ominously, at the instant when L_b occurs, Marcus will say that L_f is also happening. Both events happen at the same instant and are fully real for Marcus. For Andrew, however, when L_b occurs, L_f is still in his future. If Andrew is a presentist, he does not take L_f to be real: the future does not exist. Yet for Marcus, L_f is fully real at the same instant as L_b. Which events count as real will be different for the two observers.

Things only get worse at larger scales. Following a famous example by Roger Penrose, say instead that Marcus and Andrew are walking past each other on the street and both say "now" just as they pass. What events are happening simultaneously in the Andromeda galaxy with respect to their two reference frames? The difference between them is striking. Which events count as "now" for Marcus and Andrew can vary by several days. "For one of the people, [a] space fleet launched with the intent to wipe out life on the planet Earth is already on its way; while for the other, the very decision about whether or not to launch that fleet has not yet even been made!" (Penrose 1989, 201). This conclusion is based on the same reasoning as the train example. What events count as "now" vary from one reference frame to another. And since no reference frame can be considered "right" or "the true inertial frame," there is no fact of the matter about whether the decision to launch is in the past, present, or future.[7]

The problem for presentism, following Hilary Putnam's (1967), is that we are caught between two intuitive principles. First, within my reference frame, everything happening simultaneously with events here and now should count as real. Second, I should say the same thing for anyone else here regardless of their reference frame. The first principle says that if the present is real, then all events in the present, no matter how far away, should likewise count as real. That seems right. Events happening now exist regardless of how far away they might be. The second principle says that my reference frame is not special. The person speeding by me as I get the mail is in a different inertial frame, but if something exists for him then I should grant that it exists for me. Again, this isn't a matter of differing perceptions, but of what actually exists in different reference frames. The second principle says that ontology ought not depend on matters of transportation. The problem is that some events will be simultaneous for the person in the car when it goes by, and therefore fully real for that person, that are in my future with respect to my reference frame. But whatever is real to the driver should also count as real for me. Hence, I must consider events that are in my future as fully real, which the presentist cannot do. The same goes for Marcus and Andrew in the Andromeda example. The invasion is in the present and therefore real for Marcus. Andrew must also then count the invasion as real even though by his lights that event is in the future.

Things only get worse if we run the thought experiment the other way. Will I get a new car within the next year? That would seem to depend on my circumstances and choices between now and then. From the point of view of a distant alien, however, his "now" (or more technically, his plane of simultaneity) includes Earth in precisely 365 days and hence is fully real to him. There is a fact of the matter in the alien's reference frame about whether I still have my beat-up Oldsmobile or not. But if no reference frame can be considered privileged, I should count his reality as fully real. Hence, there is a fact of the matter regarding my future car. Given the limitations on the speed of light, the alien could never know that truth about my car and then communicate it to me in advance, but that's beside the point. The question has to do with whether events in my future already exist. If STR forces us to say that future events exist, then presentism conflicts with one of the greatest discoveries of contemporary physics—an unenviable position for any philosophical doctrine.

If you weren't familiar with the relativity of simultaneity entailed by STR, all this might still seem a bit puzzling. A few years after Einstein's publication of STR, mathematician Hermann Minkowski set the theory into a more intuitive, geometrical framework. Relativity isn't so difficult, it turns out. All you have to do is stop thinking in three-dimensional (3D) terms! This will be important when sorting out presentist responses to relativity.

3.2.3   Minkowski Spacetime

Besides the issue of simultaneity and reference frames already mentioned, there are other well-known phenomena associated with STR, namely, length contraction, time dilation, and mass variation. Going back to the train example, from Andrew's point of view, a clock in Marcus's train car seems to move too slowly. (Actually, the train would have to be traveling at more than half the speed of light for it to become readily noticed using standard clocks.) In addition, Marcus's train car and everything in it seem a bit squeezed to Andrew compared to the same train when it was in the station. And if Marcus were to toss an apple out the window, the impact it would have from Andrew's point of view would suggest that the mass of the apple was more than it should be. Each of these effects of STR is experimentally confirmed,[8] but why do they happen?

We were taught in school that our world is welldescribed by 3D Euclidean geometry. The now standard view in physics, first presented by Minkowski, is different. STR entails that reality is four-dimensional (4D). Time is the 4th dimension of a single entity: spacetime. Strictly speaking, there is no such thing as time alone or space alone. Time and space are part of a single, irreducible structure.

When physicists describe this 4D world,[9] they usually start with a metric: a map of distances between spacetime points. In 3D Euclidean geometry, the distance between two points is given by the Pythagorean theorem

$$s^2 = \left(x_2 - x_1\right)^2 + \left(y_2 - y_1\right)^2 + \left(z_2 - z_1\right)^2$$

In 4D Minkowski spacetime, the metric is given by

$$s^2 = \left(x_2 - x_1\right)^2 + \left(y_2 - y_1\right)^2 + \left(z_2 - z_1\right)^2 - \left(t_2 - t_1\right)^2$$

While the observed time between events varies from one observer to the next (time dilation) and the observed lengths of objects also varies (length contraction), the interval, *s*, is the same for everyone. The fact that the interval is invariant for all observers is the main reason for thinking of *it* as real rather than conflicting reports about length, duration, and mass.

Using the more precise differential notation,[10] the metric becomes

$$ds^2 = dx^2 + dy^2 + dz^2 - dt^2$$

The last term is negative because spacetime is not fully Euclidean. Unlike moving from a two-dimensional (2D) square to a 3D cube, spacetime models do more than simply add another dimension. Intervals in spacetime can be negative. (Mathematically, this has to do with the speed of light being constant.)

Taking spacetime realistically means there is no such thing as length itself or temporal duration itself. Observers in different reference frames will disagree about those. Strictly speaking, the only facts of the matter are spacetime facts, principally facts about the interval between events. The interval is invariant across reference frames. On Minkowski's view, the odd effects of relativity—length contraction and the like—are the consequences of experiencing our 4D reality from different 3D perspectives. Our experience of time and space is merely the 3D shadow cast by a 4D spacetime.

3.2.3.1   *Consider an Analogy*

Say we set a coffee mug in the center of a classroom and give every student a camera. Some students are seated, some are standing, and some are up on their chairs. Each one takes a picture. Naturally, we will see rather different 2D snapshots depending on the angle of the camera. Some show a handle; others do not. Some show a logo; some do not. No one photo is *the* right picture. They are just different perspectives of a 3D object in 2 dimensions. The same goes for two observers moving relative to each other. Their measurements differ because they are viewing an instantaneous 3D snapshot of a 4D reality. No reference frame is privileged in the sense that no one has the right measurement, just the way no one student has the right picture. The same goes for time and mass measurements. The only facts of the matter are spacetime facts—facts whose truth is grounded in the invariant 4D metric.

Treating spacetime as a 4D manifold with a metric is sometimes referred to as the "spatializing of time." Just as the whole of space presently exists, so does the whole of time. Again, space and time in STR are considered dimensions of a single entity. Talking about it using our normal, tensed language gets confusing. To say that the whole of time *presently* exists isn't quite right since the present has to do with an instant of time. To say that the past and future exist is to take a God's-eye point of view. If one could see reality as it is, one would find the entire block universe: all of (what we think of as) the past and future plus the whole of space in a single 4D manifold. But while the whole of spacetime is "out there," there is nothing within it that represents the passage of time itself. Time does not pass or flow in the block universe any more than space does. The whole of time exists along with the 3 dimensions of space. Nothing within the block universe represents a moving "now." "In the Minkowski world, there is no change since the whole history in time of a three-dimensional body is entirely given as the body's four-dimensional worldtube" (Petkov 2009, 163).[11]

In sum, taking 4D spacetime at face value seems to imply the following:

- Our subjective experience of the passage of time is an illusion.
- There is no present, no special ontological reality to the "now."
- What we think of as the past and future are metaphysically onapar.

That alone is enough for concern. If the standard view of relativity is correct, then our commonsense view of time is radically mistaken. As philosopher of physics Tim Maudlin says, "The moral of the theories of relativity is not that classical spatio-temporal notions are rendered merely relative, but that they are expunged from physics altogether" (Maudlin 2008, 156). Let's now connect the dots to free will.

3.2.4   The Block Universe and Free Will

If time is a static dimension within a 4D reality, then one's life is something like the music on a compact disc. All of the music is there on the disc even though we can only experience one instant at a time. More importantly, once the disc is created, there's no changing it. One's timeline is a fixed part of the metaphysical landscape. This isn't precisely *determinism* as it is usually understood, but it's a close cousin. Consider the causal determinism of a billiard ball. The ball must move the way it does because of the impressed forces. It cannot do otherwise given the laws of nature. But notice that someone like Laplace, who believed in causal determinism, would likely have been a presentist regarding time. He believed that the future must evolve in whatever way Newton's laws dictate, but that future does not yet exist. In other words, Laplace believed the future is determined but not yet fixed.

When it comes to eternalism, philosopher Joseph Diekemper points out the issue isn't determinism so much as fixity (2007, 432). What we think of as the future is no different than the past in a block universe. So in whatever sense one considers the past to be fixed also applies to the future. If it is impossible to change one, it is impossible to change the other. Consider a game of poker. Say that the deck is shuffled and dealing is about to begin. While one may speak of the cards that might turn up in one's hand, the "might" here merely indicates a lack of knowledge. The order of the cards in the deck is fixed even though the players can't yet see what they are. The cards haven't yet been revealed, but there is a fact of the matter regarding which cards will be dealt. Similarly, an eternalist might talk about different "possibilities" in the future, but every event in the timeline has precisely the same ontological weight. It is not the case that different possibilities are being made actual as time progresses, to put an Aristotelian gloss on it. Time does not progress in a B-series. Future events are metaphysically fixed, just like the past.

The problem is compounded for the theist. If eternalism and theism are true, then there is a spatiotemporal fact of the matter "out there" about what car I will be driving in 2021. Since God is omniscient, he knows all of the truths in and about physical reality. If x is ontologically real, then God knows everything there is to know about x. If God believes that I'll still be driving my beat-up Oldsmobile 10 years from now (let us hope not), then you can bet that's the car I'll be driving in 10 years.[12]

There are two plausible ways around this problem. The first is to reject libertarian freedom and settle for *compatibilism*. Free will according to the compatibilist is not what the libertarian says it is. It's not about being able to choose something other than what one actually chose. Freedom is merely a lack of coercion or constraint. So long as no one forces you to choose one option over another, you're free. For example, if I'm tied to a chair, I'm no longer free to lift my arms. But if one is acting in accord with one's internal desires, that person is free. The compatibilist also believes that determinism is correct in that deep down, our internal desires are determined by biological causes and the laws of nature. But we're still free, says the compatibilist. Determinism and freedom can coexist. If an act is done voluntarily, that is, according to one's strongest desires, it's free, even though one cannot ignore those desires and choose something else.

Compatibilists have no problem with eternalism. So long as each person at a given point in time is acting according to his or her own desires, then compatibilism is fulfilled. The compatibilist is okay with good old fashioned determinism at a deep level, so the idea that the future is fixed is not an issue.

Libertarians object that compatibilist freedom is a poor substitute for the real thing. They generally refer to compatibilism as *soft determinism*. "Soft" in that compatibilists still believe in some form of free will; "determinism" because they believe that determinism is true but hidden in the deep recesses of chemistry and physics.

A second reply to the problem of a fixed future is to distinguish what *will* happen in the future from what *must* happen. To say that I will be driving the Oldsmobile in 10 years is not to say that I will necessarily be driving the Olds in 10 years. In other words, eternalism does not imply fatalism (x must occur). In Greek mythology, the Fates proclaimed that the son of Thestias would die when a log in her fire was consumed. So even though the log could be hidden and plans laid to keep it from burning, no matter what Thestias and her son did, reality conspired against him. There was no way to avoid the fate of dying when the log was burned up.

With the distinction in hand between what must happen and what will happen, the eternalist argues that he is committed only to the latter. Yes, there is a fact of the matter about what car I will be driving in 10 years, but this fact does not entail that I must make one choice rather than another. Nothing is forcing my hand, as it were.

The libertarian can grant that events in the future are not necessary, but let's be clear about what that means. Philosophers talk about necessity and contingency using possible world semantics. In the actual world, I had a bagel for breakfast. To say that I could have had cereal is translated as 'there is a nearby possible world in which I had cereal for breakfast.' In a more distant possible world, Arthur the family cat escaped out the back door this morning to do some business in the neighbor's mulch. After rescuing Arthur from an angry octogenarian, I skipped breakfast. At a greater distance still, my mother and father never married and so I don't exist in that possible world. If a statement p is possibly true ($\Diamond p$), then it is true in some but not all possible worlds. When a statement q is necessarily true ($\Box q$), then it is true in all possible worlds.

Back now to the nonnecessity of the future. Even if eternalism is true and the future is fixed, there is a *sense* in which I could be driving a Honda Civic or—somewhat less likely—a Lexus 10 years from now: there is a possible world in which I will own a Lexus. The future is fixed in that world, but in a way that is different than the actual world.

The question is whether any of this provides room for libertarian freedom within an eternalist framework. If it does, I don't see how. It's all well and good that the Holocaust never happens in another possible world, but it did happen in the actual world and nothing can change that. If the future is just as fixed as the past, there doesn't seem to be a place for libertarian freedom, as Diekemper explains:

> [All] this means is that in the actual world, though the future is not fixed *necessarily*, it is fixed contingently. The past is also thus contingently fixed, and yet this fact does not give us any comfort when we find ourselves regretting a past event. It is not as if someone could console us by saying, 'Just because the past is actual, doesn't mean it is necessary. Things *might* have gone differently.' The appropriate reply would be, 'But they didn't! And the occurrence of that event is now inexorable.' (Diekemper 2007, 439)

If one is on a rollercoaster that is about to make a hard left turn, the eternalist is right to say that the car doesn't *have* to turn left (i.e., it's not necessarily the case). It *could* turn right in the sense that it *does* turn right in a different possible world. But, as the presentist will rightly point out, in the actual world it turns left and so that's where the car is going to go. To say that the rollercoaster might have been built differently is fine as far as it goes. But if there is only one timeline for the actual world that completely and unchangingly exists, it just isn't clear where room for free choice is to be found. Granted, events in the future are not necessary. Neither are events in the past, but we don't have the power to change the past because it is fixed. If eternalism is true, the future is fixed in precisely the same way. Too much of the literature on this issue turns on debates about necessity. In my view, that is not what the libertarian is worried about. The issue is the fixedness of the future, not its necessity.

"But that's *not* the issue," some will reply. "So long as I am the one forging each bit of the chain, the existence of the entire chain is not in conflict with free will." Here, we find a divide even among libertarians. The crux of the issue is whether libertarians need to accept the principle of alternate possibilities (PAP) which says that for one to be free, one must be able to do otherwise when he/she acts. Freedom requires that the future have live options from which a person can choose, on PAP, rather than one and only one path. A choice changes the trajectory of the future. Libertarians like Linda Zagzebski (2000) and Kevin Timpe (2008) have argued against PAP. Timpe has said that he is not terribly concerned about the metaphysics of time (private conversation). Libertarian freedom is about the source of one's decisions: are my decisions really *my* decisions or are they somehow due to other forces? So long as it's the former, I am free. The existence of futurefacts is not a problem, he believes, if the agent's undetermined will is the cause of his actions.

Most libertarians believe this is not enough. They would argue that PAP is at the core of freedom and that the ability to change the trajectory of events is a necessary condition for free will. Note that this is not exactly the same as "changing the future." I have a book on my shelf by Hermann von Helmholtz that I haven't touched in years. Unless I choose otherwise, it will remain at that very spot for another year. But I can change how the future will unfold: I can decide now to move that book elsewhere. What would have happened is not going to happen because of my free choice. In a block universe on the other hand, there is only one trajectory of events, and like the rest of the timeline, it is fixed. What we call the future exists without change. Whatever the spatiotemporal facts are about my Helmholtz book, they are what they are and nothing I do now can alter those facts. While not a direct contradiction of PAP, it is at the very least difficult to see how eternalism allows space for PAP. Hence, whether the future is open or fixed is precisely the issue.

That eternalism conflicts with free will is, admittedly, a more contentious claim than that it conflicts with the flow of time. Nonetheless, if libertarian freedom is compatible with eternalism, I have yet to see a convincing argument. My intuitions on this follow philosopher Michael Lockwood who argues that the block universe of Minkowski spacetime "rules out any conception of free will that pictures human agents, through their choices, as selectively conferring actuality on what are initially only potentialities" (2005, 69). Many eternalists agree. "In the Minkowski four-dimensional world … there is no free will, since the entire history of every object is realized and given once and for all as the object's worldtube" (Petkov 2009, 172). Minkowski spacetime therefore appears to imply eternalism and three additional theses:

- What we think of as the future is fixed.
- From a God's-eye point of view, the block universe is static and one's choices cannot alter it.
- Without the ability to influence the trajectory of events, there is no libertarian free will.

For many theists, this is an unhappy state of affairs. In particular, it isn't clear how creatures frozen within the block universe can have any sort of interpersonal relationship with God. Nor is it clear how one can be morally responsible before God if one cannot change course—*repent*, to use a more theological term—and choose to obey God's commands.

Again, not all theists are libertarians, and some libertarians are okay with the notion of an actually existing future of some kind or other. Most theists are not okay, however, with the claim that the passage of time is an illusion. In particular, the theology of intercessory prayer would have to be completely reconsidered under eternalism. If physical reality is temporally static, then God cannot consider a prayer at time t_1 and then do anything to influence events at t_2. Whatever facts of the matter obtain at t_2 cannot be changed. At best, God's "answers" to prayer would have to be preloaded into reality at the creation, which was Aquinas's view. Eternalism also has implications for what we think of as events in the past, even death. When Einstein's friend Michael Besso died in 1955, he sent Besso's wife a letter of condolences. There, Einstein explained that while from a physical point of view their timelines no longer overlapped, Besso exists just as fully as anyone else. "Now he has departed from this strange world a little ahead of me. That means nothing. People like us, believers in physics, know that the distinction between past, present, and future is only a stubbornly persistent illusion."[13] Perhaps this provided some (little) comfort to the widow, but this view also entails that Nero, Hitler, and every other evil person in history exists just as fully as you do at this moment.

In short, most theists will find *something* worrisome here about the implications of spacetime and the block universe interpretation. Let's now consider some options.

3.3   *Contra* the Block Universe

Given the way textbooks are written, one might think that what has been presented here just is the scientific view. "If you don't like the implications of spacetime, then you are free to close your eyes and ignore it, but 4D is the structure of reality according to physics." As you might have guessed, things are not nearly that simple.[14]

3.3.1   *Relativizing the Present*

Presentism says that only those things that exist now are real. Philosopher Howard Stein (1968) shows how STR can be taken one step further, fully embracing the relativity of the present.[15] Instead of expanding our ontology from 3D to 4D, reality should instead be shrunk down to a single, moving point for each observer. Here's the idea.

The place I think of as *here* and the time I think of as *now* is different from your *here–now*. Let my *here–now* be point O in Figure 3.3, a light-cone diagram (one spatial dimension is suppressed). The past for O is constituted by all the events that could (moving at the speed of light) causally influence the *here–now*, those in the past light cone. Events outside of the cone, "spacelike" separated from O, cannot influence events inside the cone without moving faster than the speed of light, which is physically impossible according to STR. Causes originating from inside the cone, "timelike" separated from O, can influence O. The fundamental difference between spacelike and timelike separated points shows that time is not just another dimension on a par with the three spatial dimensions. If it were, one could just as well orient the light cone along the x-axis rather than the t-axis, but that wouldn't make any physical sense. Causal signals, including light, only propagate from $-t$ to $+t$. The apex of the cone therefore represents a real, physical distinction between past and future with respect to O.

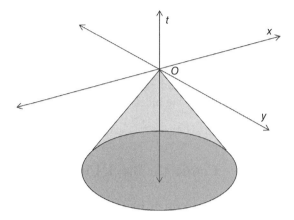

Figure 3.3   Past light cone.

Eternalism says that reality extends across the whole of space and time. Presentism says that reality extends across the whole of space at a specific time, the present. Stein takes this one step further: reality extends only to specific places at specific times, namely, the *here–now* of each observer. The past for each *here–now* point is captured by its past light cone. (The future has its own lightcone along the positive *t*-axis, but that's not important here.) Light-cone diagrams represent the metaphysical difference, argues Stein, between past, present, and future with respect to *O*. What STR shows is that, except for observers standing right next to each other, we each have our own light cones and different *here–nows*. So although the passage of time is real, Stein shows how the transition from past to present is relative to each observer, each *here–now*. Your present and mine are not the same. Note that it is not merely the perception of time that is relative on this view. The reality of past, present, and future varies from one observer to the next.

Stein takes cases like Penrose's Andromeda invaders, on the other hand, to be merely mathematical abstractions. Spacelike separated points have no causal relation to the *here–now* since they are outside of the past light cone. Therefore, the claim that a far distant event "must be 'real' or 'determinate' to an observer" at the *here–now* is baseless (Stein 1968, 16). There is no sense in which the Andromeda invasion can be real for Andrew but not for Marcus since the invasion is outside of both of their past light cones. Such examples, he says, confuse temporal calculations for time itself.

In Stein's proposal, each observer's *here–now* constitutes its own present, and each *here–now* has its own determinate past. There are potentially as many "presents" as there are observers, none of which need agree. The upside for the presentist is that Stein's view provides a real difference between past, present, and future. This "real difference" varies from observer to observer, but it is nonetheless real.

For many, however, the notion that time is completely observerdependent is a metaphysical bridge too far. Widespread disagreement among observers typically involves matters of subjective experience rather than objective reality. The perceived color of a marble might vary from person to person especially under different lighting conditions, but the mass of the marble is the same for everyone. Mass is an observer-invariant primary property. Color is an observer-dependent secondary property. Mass is important in physics, color much less so. In general, observer independence is associated with physical reality; observer dependence is associated with mere subjective experience. This is a problem for Stein's proposal. While he wants to give metaphysical weight to each observer's lightcone, making their individual *here–nows* the point of transition from past to present, the disagreement between observers is the earmark of subjective experience rather than physical reality. Hence, this view of the nature of time is not a popular view.

Instead of the present varying from observer to observer, a slightly less radical alternative is for the present to vary from reference frame to reference frame (Craig 2001, 81). In STR, all observers at rest with respect to each other constitute one reference frame no matter how far apart they are. On this alternative, everyone in a reference frame shares the same present. Past and present are still relative, but they are relative to entire reference frames rather than spacetime points. The present would differ, once again, from frame to frame; there is no universal *now*. While this view is more in the spirit of Einstein's original (pre-Minkowski) version of STR with its emphasis on inertial frames rather than spacetime points, its downside is the same as Stein's pointwise view: there are many different "presents" and so many different realities that vary from frame to frame. Again, widespread disagreement between observers generally indicates subjective experience rather than physical reality.

The bottom line is that relativizing the present to either reference frames or spacetime points does indeed avoid eternalism, but neither alternative is attractive.

3.3.2   *Manifold Antirealism*

Recall once more the two principles that Einstein was trying to reconcile via STR: (1) the laws of nature work the same within every inertial frame, and (2) the speed of light is invariant for all observers. Standard introductions to STR start with these two principles and then move to a discussion of light signals being bounced from moving rods and synchronized clocks. If all inertial frames are empirically equivalent (experiments within one have the same results as those in any other) and the speed of light is the same for all, some unusual observations will follow, namely, time dilation, length contraction, and mass variation. Clocks appear to run too slowly, objects seem too short, and masses appear to get larger when they are moving relative to us. Some data will be agreed upon by all, such as the interval *s*. Other observations, such as whether two events are simultaneous, differ from one reference frame to another.

At the beginning of his career, Einstein was focused on epistemic questions: What can an observer in one reference frame infer about the goings on in another frame by way of information traveling at the speed of light? Under the reigning philosophy at the time, known as *positivism*, metaphysical issues were a distraction at best. Positivists such as Ernst Mach taught that every part of a legitimate scientific theory must be directly supported by experience. Unobservable entities were frowned upon. (Note the metatheoretic shaping principle here.) The young Einstein was heavily influenced by Mach, and he formulated STR in accordance with positivist principles.

Einstein's positivism and the fact that Minkowski's view was developed 2 years after the publication of STR suggests that relativity need not be wedded to spacetime. Einstein himself developed STR without believing in the existence of a 4D manifold—a theoretical entity. Like point particles and ray optics, many physicists today take spacetime as a heuristic device or model for dealing with Lorentz transformations (STR) and Einstein's field equations (GTR). This is not unusual. Models and idealizations are commonplace in physics. No one literally believes that fluids are perfect continua or that condensed matter is held together by a lattice. Nonetheless, thinking of physical systems in these ways can be quite useful. What had been an intractable mathematical problem becomes manageable.[16] Following this line, spacetime can be interpreted in an antirealist way: it is useful but it does not literally exist. As Robert Geroch puts it, "If you like, 'four dimensions' is just a convenient way of describing the world and thinking about the world, nothing more" (1978, 12). Reifying spacetime is bad metaphysics, as philosopher Max Black argues:

> [This] picture of a "block universe," composed of a timeless web of "world-lines" in a four-dimensional space, however strongly suggested by the theory of relativity, is a piece of gratuitous metaphysics…. Here, as so often in the philosophy of science, a useful limitation in the form of representation is mistaken for a deficiency in the universe. (1962, 181)

While everyone agrees that the mathematical structure of spacetime has tremendous utility in physics, that does not entail that 4D spacetime exists.

Some go even further than Black, believing that this utility has become intoxicating. Nobel laureate Steven Weinberg argues that an overemphasis on geometrical explanations in spacetime models has actually hindered the development of fundamental physics (1972, vii). First, a bit of background. According to GTR, gravity is not a force. It is instead a phenomenon caused by the distorting effects of objects in spacetime. When a light ray comes near a star, it no longer moves in a straight line. The mass of the star distorts spacetime so that the light bends toward the center of mass. Light travels in the straightest possible path, but paths in spacetime(*geodesics*) are warped by objects with mass. This warping is not caused by gravity; the warping is gravity. Gravity is a geometrical property of spacetime that makes objects appear as if they were pulled by an invisible force. Relativistic gravity does not literally reach out and connect one mass to another. Instead, the mass of the sun distorts spacetime so that planets travel in a kind of depression. When a planet travels along its orbit, it is more like a toy car traveling along a track rather than a kite on a string. Orbits are caused by the effects of mass on spacetime, not the action at a distance of Newtonian gravity.

So what's Weinberg fussing about? It is well known that this geometric interpretation of gravity is inconsistent with the Standard Model of particle physics. Instead of seeing gravity as part of the geometry of spacetime, the Standard Model explains the four fundamental forces in terms of an exchange of particles. These two approaches conflict. Gravity cannot be both an exchange of discrete particles and the curvature of a continuous spacetime. When graduate students are given one picture in their courses on gravitation and a completely different picture in particle physics, Weinberg thinks there is an unnecessary muddying of the waters. He believes that the particle view of gravity will 1 day win out. In the meantime, the GTR picture of gravity hinders physicists from thinking about the problem in the right way. Weinberg would have them keep in mind that spacetime is merely a "mathematical tool," nothing more.

A closely related problem for spacetime is that, unlike STR, quantum mechanics allows for absolute simultaneity of events. Particles that are *entangled*—a technical term—have interdependent states even when the particles are spatially separated. A measurement performed on one entangled particle has an instantaneous effect on the others no matter how far apart they are. The events are simultaneous according to quantum theory, and this simultaneity is completely independent of how observers are moving. Unlike STR, quantum mechanics entails that there is an objective fact of the matter about simultaneity regardless of whether observers agree about it. If this is correct, then something must be wrong with the notion of time presented by STR which explicitly denies observer-independent facts about the sequence of events.[17]

All this shows that physics itself provides reasons for interpreting spacetime in less than realistic terms. First, it is not compatible with our current understanding of particle physics, which is Weinberg's point, or with the absolute simultaneity entailed by quantum mechanics. Second, while idealized mathematical structures in science are undeniably useful, one must draw metaphysical inferences with care. Separating the idealizations and artifacts from the realistic bits of physics is not always easy, as Larry Sklar cautions:

> While our total world-view must, of course, be consistent with our best available scientific theories, it is a great mistake to read off a metaphysics superficially from the theory's overt appearance, and an even graver mistake to neglect the fact that metaphysical presuppositions have gone into the formulation of the theory, as it is usually framed, in the first place. (Sklar 1981, 131)

This is why presentists have not simply capitulated to eternalism, STR, and spacetime. The metatheoretic shaping principles employed in the creation of a theory can be scrutinized just as much as the theory itself. Even then, one must take care in drawing metaphysical implications from "the theory's overt appearance." How things seem is not always how things are.

3.3.3   *Before STR: Lorentzian Mechanics*

As I mentioned in the previous section, Minkowski developed his approach 2 years after the publication of STR. Spacetime is an interpretation of relativistic effects. STR does not logically entail the existence of spacetime.[18] So what if one wants to be a scientific realist about relativity, but not accept the spacetime interpretation? To answer that question, we must first consider a famous anomaly.

The Michelson–Morley experiment was one of the great failures in the history of science. It started with the commonsense idea that for there to be a wave, there must be some sort of stuff for the wave to move through. Without a material medium—water, air, etc.—there can't be a wave. In the late 1800s, it was taken for granted that light waves also needed a medium in which to move: the luminiferous aether. However, the Michelson–Morley experiment[19] failed to detect how the speed of light was affected when a light source is moving with or against this aether. The speed of all other waves depends on whether they are moving upstream or downstream within their medium. The speed of light does not.

Physicists such as Hendrik Lorentz took a creative route to explain this anomaly. They held that movement through the aether causes material objects to contract, including the apparatus used in the experiment. The contraction of the equipment meant that light moving against the aether had a shorter distance to travel. Lorentz proposed that this contraction precisely cancels out the influence of the aether on the speed of light. The faster the equipment has to move through the aether, the more it contracts, and this contraction perfectly compensates for the slowing of light moving in the same direction. This influence on experimental devices made it impossible to detect how much light was being retarded by moving against the invisible "aether wind." It's as if two identical boats were released from the same place moving full speed on a river, one upstream and the other across the current. Normally, they shouldn't get back to the starting point at the same time since the boats are affected by the current in different ways. But what if the distances were changed to compensate for the current? We give the slower boat a shorter distance to travel so that both boats arrive back at the same time.

That is precisely what nature does in the case of light moving through the aether, said Lorentz. The beam of light being held back by the aether wind has a shorter distance to travel because the lab equipment itself contracts in the face of that wind. Light *seems* to travel at the same speed in every direction because of the compensating effects of this contraction. Lorentz believed that movement through the aether causes length contraction, a real physical effect. Unlike what Einstein would conclude, Lorentz held that there *is* a special reference frame in nature: the one at rest with respect to the universal aether. Because of its effects on material objects, including lab equipment, there is unfortunately no way for us to detect our velocity relative to this preferred frame. There is a truth regarding how fast each object in the universe is moving, but there is no way to find it.

There is one seemingly fatal problem for Lorentzian mechanics: there is no luminiferous aether. Like caloric and phlogiston, aether became yet another theoretical entity of the eighteenth century that wound up on the trash heap of science. The idea that movement through aether induces the contraction of lab equipment is often held up as a classic *ad hoc* explanation for the unexpected failure of the Michelson–Morley experiments.

Modern neo-Lorentzians think otherwise.[20] In their view, Lorentz was right that not all inertial frames are equal. Although the aether does not exist, there is a fundamental reference frame in nature. Under a neo-Lorentzian interpretation of relativity, time dilation and length contraction occur when one is moving relative to this preferred frame. These phenomena are not merely the result of seeing different 3D slices of a 4D reality, as orthodox spacetime theorists hold. They are instead real physical effects on 3D objects moving relative to the universal reference frame. On this view, reality is not fundamentally 4D. The correct model would be more Newtonian, 3 + 1 (3 dimensions of space plus an independent dimension for time).

It is widely recognized that neo-Lorentzian relativity is empirically equivalent to STR.[21] There are no experiments that can disprove one in favor of the other. So why is the spacetime interpretation so dominant?

One reason is the impossibility of detecting motion relative to the fundamental reference frame. Just as Michelson and Morley were unable to measure changes in the speed of light through the aether, we have no way of detecting the neo-Lorentzian fundamental frame. The physical effects of motion relative to the fundamental frame—length contraction and the rest—corrupt any experiment one might use to measure such motion. Nonetheless, neo-Lorentzians believe that there is a fact of the matter regarding our absolute motion relative to the fundamental frame. It is a fact that experiments cannot detect.

The second and more important problem for Lorentzian mechanics is that while it is empirically equivalent to STR, matters are different in GTR. STR is an idealized, special case of GTR. Once gravity is introduced, one must use covariant field equations. Such equations are needed to accommodate the fact that the distribution of mass and energy changes in GTR along with the geometry of spacetime. This is important since the equations governing GTR cannot be so easily deconstructed from 4 dimensions into

3 + 1.[22] The time coordinate is inextricably bound up with space in GTR. So while Einstein did not need the concept of spacetime in order to discover STR, it seems to be essential for GTR.[23] A space-plus-time (3 + 1) geometry can be distilled from STR in ways that GTR will generally not allow.

The upshot is that while a 3 + 1 neo-Lorentzian alternative to STR is plausible, more is needed for general relativity. If so, then in order to avoid a block universe, the neo-Lorentzian will need either to retreat to manifold antirealism (see the previous section) or look for some new structure to serve as a surrogate for time. What that structure might look like is the subject of the next section.

3.3.4   GTR and Cosmic Time

Some would argue that the attention given to STR here is misplaced. As we've seen, STR is itself an idealization that ignores gravity. As such, "it cannot be viewed …as a fundamental physical theory" (Dorato 2002, 254–255). If we're trying to discover what physics says about a metaphysical question, we should look to the more realistic and fundamental theories, like GTR and quantum mechanics, rather than less realistic and more idealized theories, such as STR and classical mechanics.[24] As it turns out, GTR offers footholds for the presentist not found in STR.

A common way to reintroduce a dynamic notion of time into GTR is to slice or *foliate* spacetime, like slicing a salami (Figure 3.4). Time can then be considered the passage from one slice to the next. For many of the standard solutions to Einstein's field equations (known as FLRW models), natural symmetries emerge along which a foliation might be made. This sequence of slices allows one to define *cosmic time*: a measure of the duration of the universe itself. This is the notion of time used when cosmologists tell us that approximately 14 billion years have elapsed since the Big Bang.

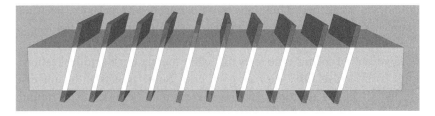

Figure 3.4   Foliated spacetime.

Let's consider a few details. In order to simplify these models, cosmologists use idealizations similar to those in fluid mechanics. The matter in fluids, at least at the scale we interact with them, can usually be treated as continuous rather than atomic. That's how engineers think of air when designing planes or cars. Likewise, on a very large scale, space itself can be treated either like a fluid or a homogeneous dust in which whole galaxies are reduced to particles. In an expanding fluid/dust universe, one can define a "fundamental observer" who is at rest with respect to the matter in his vicinity. For such an observer, the universe is isotropic, that is, it looks the same in every direction. In an expanding universe, each fundamental observer would be able to take note of two important changes: the decrease in the average mass–energy density and the temperature of the cosmic microwave background. Fundamental observers could in principle synchronize their clocks to these processes. It is possible in GTR to then link each fundamental observer in the universe together to form a *hypersurface* through spacetime. A hypersurface is a temporal slice through spacetime and represents one instant of the local ("proper") time of each fundamental observer's clock. Once spacetime has been foliated with hypersurfaces, the hypersurfaces can be ordered in sequence to represent the passage of time as experienced by a fundamental observer. This sequence is cosmic time, a universal time that is independent of moving reference frames. Again, this is a very common way of understanding time in Big Bang cosmology.[25]

Time has been recovered from GTR, so why isn't this the end of the story? There are three reasons. The first has to do with the idealizations needed to make this all work. An old joke starts with a farmer asking his physicist friend for help with his sick chickens. After scribbling some notes, the physicist exclaims, "I've got it! Unfortunately the solution only works for perfectly spherical chickens in a vacuum." Just because things work in highly idealized models doesn't mean they work in the actual world. Critics of cosmic time point out that the Friedman–Walker models require perfect spatial homogeneity and isotropy, but our universe is not a homogeneous dust or an ideal fluid. Our universe is clumped with galaxies. So while cosmic time might apply to an FLRW world, our universe is not one of those.

Advocates of cosmic time reply that while these models are idealizations, most cosmologists think they are good ones. Some aspects of the universe are in fact highly isotropic, such as the cosmic microwave background. Thus, the FLRW models are approximately true in the sense that the exact right model is somewhere in the neighborhood.

The second objection has to do with foliations in general. Within a given model, there are many ways to slice up the block. Nature doesn't force us to choose any particular foliation scheme and each scheme will present a different order of events in the cosmos. To many, this makes foliation arbitrary. If there is such a thing as the passage of time and absolute simultaneity across a hypersurface, shouldn't that somehow present itself? Is the arbitrary imposition of slices the only way of recapturing time? With so many equally justified alternatives, say, the critics, no one set of foliations can be considered "real time." It would be better to simply take the block universe as it presents itself. Time is like so many other discoveries in the physical sciences: the manifest image is not the scientific one.

In response, advocates of cosmic time agree that foliation schemes seem arbitrary if one thinks of spacetime like a pan of JELL-O, but there is in fact more to the story. Some ways of slicing the block are far more natural than others. Philosopher of physics Roberto Torretti puts it this way:

> [Where] one [foliation] is possible, many more are available as well. Not all of them, however, will be equally significant from a physical point of view. Thus, in a universe that admits a Robertson-Walker line element, the hypersurfaces orthogonal to the worldflow of matter provide an unrivalled partition of events into natural simultaneity classes. (1996, 230)

This is a widely shared view. In the FLRW models, the trajectories (or *worldlines*) of the fundamental observers are unique. No other worldlines within these model universes are guaranteed to maintain homogeneity and isotropicality. Hence, there is something physically special about defining a time scale by means of the experienced (proper) time of such observers. By analogy, consider a stack of typewritten paper with the manuscript for this book. Let's say that the pages of the manuscript have been fused with paper-mache paste. (I can't prove it, but my youngest son, Christopher, was almost certainly involved.) There are many ways one might slice up this solid block of paper, but there is only one way that will recover the flow of chapters and page numbers. Just because there are other ways to cut the block does not mean that all such cuts are on a par.

The third objection has to do with the contingent nature of cosmic time. Presentists generally believe that time, like space, is a fundamental component of reality. It is so fundamental that presumably all nearby possible worlds would also display a passage of time, just as all nearby possible worlds share the same laws of nature.[26] However, as Kurt Gödel showed (1949, 520), this is not the case. Craig puts Gödel's point this way:

There are other cosmological models which do not involve homogeneity and isotropy and so may lack a cosmic time altogether. Cosmic time is thus not nomologically necessary, and its actual existence is an empirical question. (2001, 206–207)

Some physically possible worlds have cosmic time; many do not. Craig himself is not bothered by this lack of nomic necessity. His critics, on the other hand, want to know if time is as fundamental as presentists think: can its very existence depend on the distribution of matter and energy in a given universe (Balashov and Janssen 2003, 342)?

The answer is, it might. Craig's move is not unprecedented in debates about time and space. Strict relationalists going back to Leibniz (1646–1716 AD) have held that there is no such thing as space itself; there are only spatial relations such as "to the left of" and "is five meters from." Objects are spatially related, but space itself is not part of the relationalist's ontology. There is no possible world in which there is space but no objects. Similarly, relationalists regarding time do not believe in time itself, only in temporal relations such as "before" and "after." Events are temporally related, but there is no such *thing* as time. In a universe with no changes of state (e.g., a world consisting of a single, stable point mass), there would be no time. Thus, relationalists have always believed that space and time are contingent. They are contingent upon the existence of objects and events, respectively. To say that time is contingent on the distribution of matter and energy is therefore not as radical as it might first appear. It's a familiar move in this debate. Moreover, the idea that time is contingent fits nicely with the fine-tuning argument in the previous chapter. Almost everything in the universe as we know it depends on the FTCs having their precise values. In light of fine-tuning, it doesn't seem so surprising that time itself might depend on physical brute facts like homogeneity and isotropy.

In the end, I would be concerned if cosmic time were merely the invention of presentists for the protection of their view from spacetime eternalism. Instead, presentists have simply adopted this fully developed cosmological idea to show that GTR need not be seen as hostile to the flow of time. Cosmic time is not the only refuge for presentists within the structure of GTR, however. As we will see, there are more recent approaches for disentangling time from space and recovering a 3 + 1 scheme.

Let's consider a loose end first. One might wonder whether cosmic time with its foliated block universe does the presentist any good. After all, even with a foliation, all the points in the spacetime manifold including the ones we think of as in the future are still "out there" in some sense. Look at Figure 3.4 again. Foliation slices the salami, as it were, but all of the parts of the salami still exist. So while the foliated block captures an objective flow of time, it also seems to entail *some* measure of existence for the past and future—something the presentist does not want. Presentism says that only the "now" is fully real. It does not allow the past and future to have some lesser, shadowy sense of existence. Doesn't this force presentist advocates of cosmic time back into antirealism? It seems they would have to deny that the whole of a foliated spacetime corresponds to reality.

3.3.4.1 *Yes and No*

Yes, the presentist has to reject the orthodox 4D interpretation of spacetime, but that doesn't mean one must be a complete antirealist about GTR. Recall the discussion of phase spaces from Chapter 2 (Section 2.3.3). Every point in a phase space represents one possible state of a system, like a pendulum or atoms in a box. Curves through the phase space represent the evolution of the system from one state to the next. The trajectory in Figure 2.4, for example, represents the changing state of an ideal pendulum. At any given instant, only one point along this curve represents the state of the pendulum, and this statepoint changes continually as the pendulum swings. A scientific realist about classical mechanics would say that this phase space is an approximately true model of a pendulum, even though only one point in the space represents the system at any given instant.

This same attitude is available to the presentist vis-à-vis cosmic time. While a foliated spacetime is an approximately true model for the evolution of the universe, only one slice of that model corresponds to the 3D universe at any instant. The rest of cosmic time represents states of the universe that have existed or will exist. One need not say that past and future points in a foliated spacetime currently have any sort of metaphysical weight. Just like the pendulum, the model indicates what the future will look like; it does not says that the future exists. So while one can believe that GTR is approximately true, that doesn't mean one must believe in the literal existence of the past and future as depicted in its spacetime models.

And so the presentist has some room to maneuver even within the confines of GTR. Of course, physics did not end with Einstein. Let's move on to some more recent developments in the philosophy of time.

3.3.5 *21st-Century Physics and 3 + 1*

When we left STR in favor of GTR in the last section, the reason had to do with scientific realism. GTR is the more fundamental and realistic theory; STR is merely an idealized special case of GTR. This move toward the more fundamental does not stop with GTR, however. Monton believes that this line of thinking should be pushed one step further (2006, 2011). Since GTR is not consistent with quantum mechanics, he argues that we should look to even more fundamental theories that attempt to reconcile the two. If new theories of quantum gravity allow for a presentist view of time, then we ought not care if superseded theories like STR/GTR do not.

While Monton's point is well taken, the physics community is widely split over the correct theory of quantum gravity (Smolin 2001). It is not yet clear whether future physics will be more hospitable to presentism than STR/GTR. Nonetheless, there are some hopeful signs.

3.3.5.1   The Canonical Approach

One approach to quantum gravity seems to be exactly what the presentist was hoping to find. Standard GTR takes spacetime to be a manifold with a metric that changes over time.[27] The incorporation of quantum mechanics means, at the very least, that something in spacetime undergoes quantum fluctuations. On the canonical (or "Hamiltonian") approach, it is the metric that fluctuates. The mathematics of quantum mechanics—operators on a Hilbert space—applies to the metric rather than the manifold. That doesn't mean that the manifold is left untouched. In fact, the 4D manifold must be split back into the classical division of space and time. Space is considered the fundamental entity, not spacetime (Rickles 2008, 323). More importantly, space evolves over time, a 3 + 1 structure rather than 4D.

Although canonical quantization is a large research program (with many branches) in quantum gravity, it isn't necessary to bring quantum mechanics into the picture. The Hamiltonian version of GTR without a quantized metric has been used by physicists for decades (Misner *et al.* 1973). As Brian Pitts shows, this Hamiltonian version of GTR is a common and perfectly legitimate mathematical form of GTR, even though it is a fundamentally 3 + 1 approach (Pitts 2004).

"How can this be?" one might ask. "The same laws cast in a different form completely changes the metaphysics? Surely some experiment must favor the 4D view over 3 + 1." Actually, no. The two versions are empirically equivalent. Any data supporting the 4D spacetime approach also supports the 3 + 1 form. The fact is that having multiple forms of the same theory is not as unusual as one might think. Physical theories are often cast in different ways depending on the application. Hamiltonian mechanics was originally invented in the 1800s as a way of understanding Newton's Laws. Graduate students easily switch between these versions of classical mechanics (and others). From the earliest days of quantum mechanics, there were two different ways of approaching the math, one developed by Heisenberg and the other by Schrödinger. And engineers have long found that working with the equations for electric *fields* is much more difficult than the equivalent electromagnetic *potentials*, and so the latter are used whenever possible.

The upshot is that GTR does not logically or mathematically entail a 4D interpretation. The canonical approach yields the 3 + 1 structure presentists seem to need. So why is this any better than cosmic time? One reason is that cosmic time is only found in FLRW models with perfect homogeneity. In other words, only idealized spacetimes will allow for cosmic time. Pitts (2004) shows that the Hamiltonian approach to GTR is not limited in this way. All of the models derived from the constrained Hamiltonian equations will have absolute simultaneity which can then be the basis of an objective flow of time.[28]

Another reason is that the foliations used in cosmic time impose extra structure that mere GTR does not have. If we consider the spacetime of GTR to be the salami, cosmic time is the slicing of the salami. GTR itself does not have and does not need foliations in spacetime. By analogy, we can impose patterns on groups of stars and call them constellations, but there are no real, physical lines between those stars. Grouping stars into constellations imposes structure over and above what physically exists. From an astronomer's point of view, a constellation is just a handy convention, nothing more. Many physicists hold this same view about spacetime foliations and cosmic time. They believe in spacetime, but not in the added structure imposed by foliating it.

The 3 + 1 Hamiltonian approach avoids this problem. No structure needs to be added to reclaim the passage of time. Space is extricated from time as a consequence of the mathematics, not an *ad hoc* addition of nonphysical structure.[29]

Eternalists do not think highly of this maneuver. Earlier, we saw how presentists might take an antirealist approach toward spacetime (Section 3.3.2), but the eternalist also has an antirealist card to play and he will play it here: the constrained Hamiltonian 3 + 1 formulation of GTR is merely a convenience. Reality, says the eternalist, is described by Einstein's original 4D field equations. This is, in fact, the majority view among physicists and philosophers of physics. All sides agree that the metaphysics of the 3 + 1 and 4D views can't both be true. Most believe the right metaphysical picture to be the one painted by 4D spacetime. The Hamiltonian version of the equations is useful for probing the features of GTR, but most think we ought not adopt the 3 + 1 structure that goes along with it.

3.3.5.2   *Hořava Gravity*

Although it's always risky to appeal to cutting-edge physics, there is a potentially important theory that has not yet found its way into the philosophical literature: Hořava gravity (Merali 2009). We've already discussed how Einstein took gravity to be a geometric property of spacetime (Section 3.3.2) rather than a force and that GTR conflicts with the Standard Model of particle physics. Particle physicists, on the other hand, believe that Newton had it right: gravity *is* a force, one of the four fundamental forces in nature, although it is experimentally more elusive than the other three. Trying to reconcile quantum mechanics, the Standard Model, and GTR is the holy grail of theoretical physics today. Physicist Petr Hořava's recent attempt to do just that has been creating quite a stir, generating over 250 research papers in the 18 months after it was first published (Ananthaswamy 2010, 28).

To understand Hořava, let's go back to the discussion of symmetry breaking in Chapter 2 (Section 2.2.2). At current energy levels in the universe, we experience electromagnetism and the weak nuclear force as two distinct forces (although 'experience' isn't quite the right word, since the weak force only comes into play in the nucleus of atoms). Electromagnetic effects, on the other hand, are all around us. You wouldn't be reading this page without them. Shortly after the Big Bang, energy levels were much higher. Before the universe cooled to around 10^{15}°K, these two seemingly different forces were combined into one, the electroweak force. At a more fundamental level than conventional physics, electromagnetism and the weak force are found to be two sides of the same coin. The point is that the universe appears one way at high energies and a dramatically different way at lower temperatures. As we'll see, Hořava makes a similar move with respect to spacetime.

According to GTR, gravity is not what it seems to be. We intuitively think of gravity as a force (although our "intuitions," we should remember, have been shaped by decades of public education that don't get beyond Newton's laws). Einstein taught us that at a more fundamental level, gravitational effects are due to the warping of spacetime. Hořava wants to go deeper still. Under his proposed reconciliation of GTR and quantum mechanics, space and time are *not* folded into spacetime at very small scales. At the most fundamental level, space and time are distinct. "I'm going back to Newton's idea that time and space are not equivalent," Hořava says (Merali 2009, 18).

Spacetime is merely something that emerges from Hořava gravity at the scale and distances of ordinary life:

> In this picture, Lorentz invariance [of GTR] is only emergent at long distances, while the fundamental description of the theory is deeply nonrelativistic. At short distances, the spacetime manifold is equipped with an extra structure, of a fixed … foliation by slices of constant time. This preferred foliation of space-time… leads to an invariant notion of time…. (Hořava 2009, 25)

For a simple analogy of what Hořava has in mind, consider a mosaic. Up close, a mosaic might look like a bunch of colored tiles or, as in Figure 3.5, a collage of small pictures. One can only see the whole figure from a distance. In Hořava's theory, GTR is how things seem at large scales. If we were able to zoom in close enough, however, we would see that spacetime in fact has a 3 + 1 structure. Space and time are distinct. Spacetime is merely the way things seem when dealing with large distances, but it is not the way

Figure 3.5　Mosaic.

things really are. Without any idealizations or imposed structure, nature at its most fundamental level is temporal.

Among the advantages of Hořava gravity are first that it fits in nicely with particle physics. Gravity is in fact mediated by a particle, the graviton, just as predicted. Second, it explains away the need for "dark matter." Dark matter has been posited to explain the motions of galaxies which seem to have far more matter in them than what we can see. Many physicists believe that galaxies must contain a vast amount of matter that cannot be visibly detected, hence "dark" matter. Hořava gravity accounts for these anomalous motions without the need for exotic, undetected forms of matter. Third, it explains away the need for "dark energy." Dark energy has been posited to explain why the universe is expanding more rapidly than it should according to GTR. Hořava's theory contains a possible explanation for this phenomenon as well, although the parameters involved must be fine-tuned (Appignani *et al.* 2009, 2010).

For our purposes, the most important aspect of the theory is its extrication of time from spacetime. If Hořava is correct, then the flow of time is not merely psychological. It is part of the structure of the universe itself.

In one way or another, several of the approaches outlined in this section (Section 3.3) take an antirealist view of orthodox relativity theory. Manifold antirealism is the most straightforward in that it *rejects* the ontology of the block universe. Spacetime is considered to be an idealization with heuristic value, nothing more. Neo-Lorentzianism and Hořava gravity *replace* STR and GTR, respectively. The former embraces Lorentz's view that motion induces real effects on matter, such as length contraction. The latter replaces 4D space-time with a $3+1$ structure. Cosmic time *adds* structure to the block universe that mere GTR does not have. Proponents of cosmic time believe that nature has temporal relations that GTR fails to capture. Research continues and I don't believe there will ever be, as some late nineteenth-century scientists hoped, a "final physics" where all of the questions are answered. But for today, there are more places for a presentist to stand than philosophers sometimes claim.

This then is the current landscape of the debate over relativity and time. Let me conclude with my own views on how best to think about all this.

3.4 Two Suggestions from the Philosophy of Science

We've been looking at specific ways that presentists try to recover an objective flow of time from spacetime physics. There are also more general approaches that can be brought into play. The first has to do with the interpretation of mathematical spaces. The second is about the limited applicability of physical theories.

3.4.1   *Metaphysics and Mathematical Spaces*

We began this chapter with worries about time and free will, and some might see it as just philosophy. "Metaphysicians might lose sleep over spacetime threats to presentism and free will," one might say, "but scientists can happily ignore all this." This is true, but only in part. In fact, physicists have always had to make decisions about metaphysics and mathematical theories, which parts are merely useful and which refer to real entities and processes.

Consider two examples. When it comes to antennas, Maxwell's equations for electromagnetism have two kinds of solutions: advanced and retarded. Intuitively, the latter specifies the field created by a charge moving along the length of the antenna. As the charge oscillates, a wavefront is created and extends outward into space. For radio transmissions, the wavefront travels at the speed of light until part of it is absorbed by a receiving antenna. So far, so good. Advanced solutions, on the other hand, seem to represent a wavefront that arrives at the receiving antenna *before* the signal was sent. If the retarded solution says that it takes 3 seconds for the signal to go from the transmitting antenna to the receiver, the advanced solution says that a signal was received 3 seconds before the transmission began. Advanced solutions reverse the order of cause and effect. While engineers find both solutions useful in different contexts, only the retarded solutions are considered physically real.

My second example is a hot topic in the philosophy of physics (Maudlin 2002, 3). Students are taught that electromagnetic phenomena can be accounted for in terms of electric and magnetic *fields*. In other words, it's the **E** and **B** in Maxwell's equations that do the work. However, the mathematics of vector fields is difficult. Engineers often find it easier to work in terms of *potentials* instead. Here's what I mean. Consider the electric field produced by a single electron. The electric potential is defined by the amount of force experienced by a test charge as it approaches this electron. There is a spherical symmetry in this case: the test charge will experience the same magnitude of force no matter where it is on an imaginary spherical shell centered on the electron itself. The electric potential is the energy stored up by the test charge as it moves from one imaginary shell to another (Hayt 1981, 100–102). (Things are more complicated when the field varies

with time.) The point of all this is that no one takes electromagnetic potentials as real.[30] Fields exist; potentials are a mathematical convenience.
A Laplacian demon would have no need for potentials.

Clearly, ontological judgments are being made here. Some theoretical
entities exist; others do not. I suggest that we understand spacetime much
like we do potentials and state spaces. Let's go back to the pendulum phase
space one last time (Figure 2.4). While each point on the oval trajectory
represents a different state that the pendulum has or will have, only one
state point can represent the actual state of the pendulum at any instant.
One and only one point corresponds to the way things are in the real system
at any given time. That is what the presentist wants to say about spacetime.
Even in a spacetime model that correctly describes spatiotemporal events
in our universe, only one temporal slice of that spacetime corresponds to
what actually exists.[31] If a given spacetime model isn't divided up into
temporal slices (foliations), then the model fails to describe a property of
our universe. There's nothing wrong with that, as I'll explain in the next
subsection. No model or theory purports to capture the whole of reality.

In short, presentists arguing against the existence of a full-blown 4D
spacetime are not making an unprecedented, *ad hoc* move. They are making the same kind of metaphysical judgment found throughout the history
of physics.

3.4.2   *Idealizations and Domains of Applicability*

Most theories in physics do not apply universally, but rather to a limited
range or scale. This goes handinhand with the types of idealizations
employed. Continuum mechanics, for example, treats matter as if it were
smoothed out and continuous across a region rather than atomic.
Aerodynamics treats the airflow over a wing the same way, and these are
perfectly good idealizations for the scale at which we normally deal with
materials, especially fluids and gases. Spacetime theorists make this same
move by ignoring midscale structure. In the FLRW models discussed earlier, whole galaxies are treated as individual particles in order to achieve
cosmic homogeneity and isotropy. The actual universe, in contrast, is neither homogeneous nor isotropic. It contains nebulae and galaxies bunched
together in various ways.

The clumpiness of the universe is not the only thing these spacetime
models ignore. Physicist George Ellis argues that irreversible processes in
chemistry and biology are also glaringly omitted:

The time reversible picture of fundamental physics underlying the block universe viewpoint … does not take seriously the physics and biology of the real world but rather represents an idealised view of things which is reasonably accurate on certain (very large) scales where very simplified descriptionsare successful. (2007, 74)

While midscale science must take account of entropy and the emergence of complex systems—phenomena that are irreducibly temporal—STR and GTR

> do not apply to spacetimes including complex systems because the equations of state they assume are too simple—they do not include friction and dissipative effects, hierarchical structures, feedback effects, or the causal efficacy of information, and they do not take quantum uncertainty into account. (2007, 74)

Ellis argues that even if spacetime theories do not contain an objective flow of time, much of the rest of science cannot do without one. "The flow of time is very apparent at some scales (that of biology for example), and not apparent at others (e.g. that of classical fundamental physics)" (Ellis 2007, 52–53).

This sort of tension between models or theories at different scales is far more common in science than one might expect. Normally, such conflicts remain hidden. As long as we're dealing with a given theory at a particular scale, all is well. But trying to stitch together the whole of physics across scales leads to inconsistencies and sometimes outright contradictions. If you find this surprising, consider that no one has ever found an internally consistent version for the whole classical mechanics.[32] Trying to unify GTR with the systems that Ellis mentions is no easier.

Given the limited domains of applicability within physics itself, it should not be surprising that trying to stitch metaphysical views about time and free will together with the tensor calculus and cosmology produces inconsistencies. There are tensions between relativity on one hand and presentism and free will on the other. So? As we've seen, relativity also has tensions with much of the rest of mid- and microlevel science. While nonlinear, dissipative, and quantum systems exist, they do not fit comfortably in the lean, abstract realm of spacetime, and that's fine. This just shows that physics as a discipline is more specialized and fragmented than most people realize. But if physics itself remains a patchwork of irreconcilable theories, why should we expect complete harmony between physics and metaphysics? Even without the proposals made in Section 3.3, I see no compelling reason to ditch presentism or libertarian freedom if that's what it would take to square our beliefs with relativity. Better to patiently wait for a successor theory that (at the very least) resolves the tension between relativity and quantum mechanics.[33]

We should note that appealing to idealizations and domains of applicability in this way is a small step back from scientific realism. GTR, chaos theory, and quantum mechanics are approximately true, but cannot be pieced together into a coherent whole. As I've argued, these kinds of conflicts give the presentist some license to suspend judgment on theories that support a static view of time, but this approach is, admittedly, a two-edged sword for the theist. I have been arguing that metaphysics can sometimes ignore scientific realism when it comes to specific theories. The problem is that the metaphysical naturalist can make the same move when it comes to, say, cosmological fine-tuning. Theists want their science-based arguments to be taken seriously, but what's to stop the naturalist from saying, "Of course I take it seriously. I just interpret that bit of science in an anti-realist way." Once we allow the antirealist horses to leave the barn, it is very difficult to get them corralled just so.[34]

Unfortunately, I see no way around this if one wants to be a scientific realist of some kind or other. Not every bit of good science *can* be interpreted realistically. There are too many tensions among all the bits of "good science." The question is how and when to play the antirealist card. There are no ironclad principles of logic or science that provide an answer. (This issue will be taken up again in Chapter 7.)

The arguments in this section are little more than elaborations of Sklar's point quoted earlier, "[It] is a great mistake to read off a metaphysics superficially from the theory's overt appearance" (1981, 131). A 4D spacetime might be the best way of working with the equations of STR/GTR, but that doesn't mean that spacetime is real. In short, current physics does not entail that the future literally exists and so presentism and libertarian freedom need not slink off to the trash heap of philosophy just yet.

3.4.3   Conclusions

Having slogged through this forest, you might be wondering why we came this way in the first place. What does religion have to do with the philosophy of time, again? Quite a bit actually.

The main culprit is eternalism. The eternalist holds a static view of time: from a God's-eye perspective, there is no special passing moment that is the present. The whole of time exists including what we think of as the future. One problem this poses for theism has to do with prayer and God's action in the world. If the entire block universe timelessly exists, then God cannot literally respond to one's prayers. Responding to prayer would mean changing events within the block universe, an entity that does not change. If God wanted to respond to prayer, God would have to destroy the universe in which those prayers are offered and then create another universe with the changes God has ordained. That's not the only way to think about things, as philosophers of religion will tell you, and we could chase down all the alternatives with another chapter. Instead I will simply state that in my view there are no viable scenarios that allow for answered prayer under a static view of time.

Another reason theists have a stake in these matters has to do with libertarian free will. As we saw in Section 3.2.4, eternalism entails that what we think of as the future has the same ontological status as the past. In whatever sense the past is fixed, so is the future. Most libertarians believe that for free will to be possible, future events cannot be fixed in this way. Decisions I make now must be able to shape the trajectory of future events; otherwise, there is no free choice. Hence, the philosophy of time should matter to libertarian theists.

These are contentious issues and there is no consensus among theists on any of them. The goal here was to show that *if* one believes in an objective flow of time and/or a robust sense of free will, then STR and GTR pose a problem in that they seem to entail a block universe. As I've argued, there are several possible moves one might make in order to preserve a classical view of time. Only one thing is absolutely clear: there is no escaping the philosophy of physics when it comes to long-standing metaphysical questions about time and freedom.

Notes

1   "Given for one instant an intelligence that could comprehend all the forces by which nature is animated and the respective situation of the beings who compose it for one instant … it would embrace in the same formula the movements of the greatest bodies of the universe and those of the lightest atom; for it, nothing would be uncertain, and the future, like the past, would be present to its eye" (Laplace 1812, 4).

2 As Bradley Monton mentions (private correspondence), it is worth noting that there are deterministic versions of quantum mechanics, such as Bohm's theory. Hence, it is not certain that quantum theory has broken the back of physical determinism.

3 Just to be clear, a factofthematter is a mind-independent collection of entities or events. *Fact, state-of-affairs*, and *truth-maker* are philosophical terms referring to things in reality as distinct from mere matters of perception. There is a fac-tofthematter regarding the precise number of atoms in my coffee mug even though no one knows what that number is or would even venture a guess.

4 Which has nothing to do with the better known political philosophy by the same name.

5 Other types of A-series include the *growing block*: the past is real and is contin-ually being added to as moments in the present slip into it; the present is the leading edge of what is real. Another is the *moving spotlight*: the past and future have a shadowy, lesser grade of existence; the present is fully real and moves along the timeline.

6 Einstein did not discover STR without help. He was advancing work previously done by Hendrik Lorentz and George F. Fitzgerald who had themselves been influence by Henri Poincaré.

7 Not everyone is happy with extreme examples like this one: "The lines of simul-taneity for a given inertial frame of reference do not determine what is currently going on at distant places, but only what dates should be ascribed to them in order to make electromagnetic phenomena coherent. The Special Theory is a theory of electromagnetic radiation and determines distances and durations, and hence positions and dates, by means of light signals. As far as electromagnetic phenomena are concerned, we have no other means of telling exactly when a distant event … takes place…. The ascription of presentness, pastness, or futu-rity, to events outside the light cone is nominal rather than real, and has no bearing on their ontological status" (Lucas 2008, 284–285). "This argument hinges on the simplistic assumption that an inertial coordinate system is associ-ated with each observer, and that what the observer perceives as his or her pre-sent necessarily corresponds to the equal-time hypersurfaces of his or her inertial coordinate system. But there is no compelling reason for the adoption of such inertial coordinates systems at rest relative to the observers. The observers are not using these coordinates for the purpose for which they are intended, that is, the simplest formulation and application of the laws of mechanics" (Ohanian 2007, 94).

8 Clocks in supersonic jets run slower compared to clocks on the ground, and particles traveling close to the speed of light in supercolliders do seem more massive than the same particles at rest. The only reason the "weird" effects of STR are not usually seen is that objects move too slowly for us to notice in the everyday world.

9   2D for two-dimensional, 3D for three-dimensional, and so on.

10   According to Maxwell's theory of electromagnetism, the speed of light c is a constant and the same for all observers. Since speed is distance divided by time, in differential notation, the speed of light is $dx/dt = c$, and so $dx = cdt$. Squaring both sides and moving all terms to the left gives us $dx^2 - c^2dt^2 = 0$. This is the equation for objects moving at the speed of light. The interval between any two events is $dx^2 - c^2dt^2 = ds^2$. A slight change of units allows us to drop the c. Adding two more dimensions (y and z) yields the Minkowski interval.

11   "Worldtube" is Petkov's generalization of an STR world line, the path that an object takes through 4D spacetime.

12   This is independent of intramural battles about divine foreknowledge, Molinism, and open theism. Molinists believe that God knows the future by way of middle knowledge, not because futurefacts exist and God perceives them. Molinists, like William Lane Craig, can affirm both divine foreknowledge and presentism. What all sides can agree on is that whatever the *spatiotemporal* facts are in this possible world—past, present, or future—God knows them. According to eternalism, what we think of as future facts have the same ontological status as those at every other point in time. So if eternalism is true and God is omniscient, then God has exhaustive knowledge of the entire static timeline.

13   Translation by Brian G. Smith, with one minor change. I have rendered "Fürunsgläubige Physiker" as "People like us, believers in physics." Smith's translation downplays the religious connotation of 'gläubige,' but in context, I believe that Einstein intended it. Lockwood agrees, translating the phrase "For us devout physicists" (Lockwood 2005, 52).

14   For a recent, in-depth defense of presentism in light of relativity, see Zimmerman (2011).

15   We should note that Stein was not endorsing any particular metaphysical view, but merely demonstrating that STR does not answer the question one way or the other (1968, 20).

16   See Misner *et al.* for the different in calculations of a test charge in an electromagnetic field for 4D and 3D models (Misner *et al.* 1973, 78–79).

17   For more on this, see Maudlin (1996).

18   Einstein himself soon endorsed Minkowski's view, but that is irrelevant here. The question is whether four-dimensionalism is essential to STR.

19   More accurately, this refers to a series of experiments begun at Case Western Reserve University in 1887 and repeated in different countries.

20   *Neo*-Lorentzian because such theories do not follow Lorentz's theory that electrons interact with the aether, a view that was thoroughly refuted by quantum mechanics(Craig 2001, chap. 9). Craig gives three criteria for a theory to be neo-Lorentzian:"(i) physical objects are n-dimensional spatial entities which

endure through time; (ii) the round trip vacuum propagation of light is isotropic in a preferred (absolute) reference frame R_0 (with speed $c = 1$) and independent of the velocity of the source; and (iii) lengths contract and time rates dilate in the customary special relativistic way only for systems in motions with respect to R_0." (Craig 2001, 14).

21   While Bell (1976) is often cited for this, he was talking about pedagogy, not philosophy. Also see Callender (2000).

22   To understand one part of why the extraction of time is so difficult in GTR, consider the distance between two points in a 2D, non-Euclidean space. Simplifying somewhat, the equation would be $s^2 = ax^2 + bxy + cy^2$. (We should be using differentials and generalized coordinates here: $ds^2 = g_{11}dx_1^2 + g_{12}dx_1dx_2 + g_{21}dx_2dx_1 + g_{22}dx_2^2$.) Notice that when $a = c = 1$ and $b = 0$, this equation just becomes the familiarPythagorean theorem. The more general form of the equation can be used in either Euclidean or non-Euclidean spaces. Euclidean geometry is one of an infinite number of possibilities. For most spaces, the middle term, bxy, will not drop away. This means that cleanly separating an x-coordinate from a y-coordinate in that space is not possible and that the relation between the two coordinates is more complex than in a Euclidean plane. Moving to a relativistic spacetime, there are four coordinates $\langle x, y, z, t \rangle$. The distance between any two points in spacetime will be given by a (far) more complex version of our distance equation. In most cases, the time coordinate cannot be isolated in its own algebraic term and so one cannot separate t from the space coordinates.

23   More technically, general covariance in the governing equations was not necessary to formulate STR, but it is for GTR.

24   This is a bit overstated, as if the only important part of science is fundamental physics. Philosophers of science draw methodological and metaphysical lessons from all of natural science, not merely the most fundamental level—if there is such a thing.

25   For more on the fundamental observer and local matter distribution, see Torretti (1996, 206–207). Other well-known foliation schemes are based on the constant mean curvature of hypersurfaces within a given model. Many of the same idealizations and conditions apply, for example, that the solutions are globally hyperbolic (Lockwood 2005, 118–121).

26   Possible world talk was mentioned in Section 3.2.4. Our actual universe is one possible world. Nearby possible worlds are much like the actual world, but with slight changes. In the actual world, I had toast for breakfast. But since I could have had oatmeal, we would say there is a nearby possible world in which I did have oatmeal for breakfast. There is a farther world in which I have four children, and so on, where the distance from the actual world represents a greater degree of change. Propositions that are true in every possible world are necessarily true. For example, triangles have three sides in every possible

world. A physically possible world is one that has the same laws of nature as the actual world.

27 One can think of a manifold as something like a 2D surface or a 3D continuous fluid. A metric is the distance between points on the manifold.

28 To be precise, there are some restrictions involved in the constrained Hamiltonian approach. Eligible spacetimes must be globally hyperbolic (i.e., causally wellbehaved).

29 We should note that Pitts's view is not the only one based on the Hamiltonian 3 + 1 version of GTR. Philosopher Thomas Crisp has a related presentist proposal (2008) drawing on the work of physicist Julian Barbour (1999). While interesting, it appears that Barbour's program has been superseded by more recent developments in loop quantum gravity (Butterfield 2002).

30 Until recently, that is. Because of the Aharonov–Bohm effect, many now think otherwise (Belot 1998).

31 Critics might complain that GTR and classical—in this case Hamiltonian—mechanics are not analogous in the way this argument requires. Classical mechanics requires a demarcation between instantaneous physical states and time that GTR forbids. I agree to some extent, but would point out that the full range of laws used in classical mechanics strengthen the argument. Classical mechanics also encompasses Laplace's equation and the heat equation, both of which have very different temporal properties as compared to Hamiltonian systems (Smith 2000). My point is merely that none of the mathematical spaces associated with these equations need to exist in order for them to be explanatorily useful. The same goes for the mathematical space of GTR: spacetime.

32 The problem stems from the many equivocal ways that classical mechanics deals with particles(see Section 7.3.5 for more).

33 Monton makes a similar point in his (2011).

34 I think I stole that line from Mark Wilson. Like many of his former students, I find it hard not to borrow from Mark.

References

Ananthaswamy, Anil. 2010. "The End of Space-Time." *The New Scientist* 7 (August): 28–31.

Appignani, Corrado, Roberto Casadio, and S. Shankaranarayanan. 2010. "The Cosmological Constant and Hořava-Lifshitz Gravity." *Journal of Cosmology and Astroparticle Physics* 2010 (04): 006–006.

Balashov, Yuri, and Michel Janssen. 2003. "Presentism and Relativity." *British Journal for the Philosophy of Science* 54 (2): 327–346.

Barbour, Julian. 1999. *The End of Time: The Next Revolution in Physics*. Oxford: Oxford University Press.

Bell, John Stewart. 1976. "How to Teach Special Relativity." *Progress in Scientific Culture* 1 (2).

Belot, Gordon. 1998. "Understanding Electromagnetism." *British Journal for the Philosophy of Science* 49: 531–555.

Black, Max. 1962. "Review of the Natural Philosophy of Time, G.J. Whitrow." *Scientific American* 206 (April): 179–184.

Butterfield, Jeremy. 2002. "Review: 'The End of Time.' " *British Journal for the Philosophy of Science* 53: 289–330.

Callender, Craig. 2000. "Shedding Light on Time." *Philosophy of Science* 67 (3): s587–s599.

Craig, William Lane. 2001. *Time and the Metaphysics of Relativity*. Boston: Kluwer Academic.

Crisp, Thomas. 2008. "Presentism, Eternalism and Relativity." In *Einstein, Relativity and Absolute Simultaneity*, edited by William Lane Craig and Quentin Smith, 262–278. New York: Routledge.

Diekemper, Joseph. 2007. "B-Theory, Fixity, and Fatalism." *Noûs* 41 (3): 429–452.

Dorato, Mauro. 2002. "On Becoming, Cosmic Time and Rotating Universes." In *Time, Reality & Experience*, edited by Craig Callender, 253–276. New York: Cambridge University Press.

Ellis, George F.R. 2007. "Physics in the Real Universe: Time and Space-Time." In *Relativity and the Dimensionality of the Word*, edited by Vesselin Petkov, 49–79. Montreal: Springer.

Geroch, Robert. 1978. *General Relativity from A to B*. Chicago: University of Chicago Press.

Gödel, Kurt. 1949. "A Remark on the Relationship between Relativity Theory and Idealistic Philosophy." In *Albert Einstein: Philosopher-Scientist (Library of Living Philosophers)*, edited by Paul Schilpp, 555–562. La Salle: Open Court.

Hayt, William H. 1981. *Engineering Electromagnetics*. 4th ed. New York: McGraw-Hill.

Hořava, Petr. 2009. "Quantum Gravity at a Lifshitz Point."*Physical Review D* 79 (8): 084008.

Laplace, Pierre Simon. [1812] 1951. *A Philosophical Essay on Probabilities*. Translated by F.W. Truscott and F.L. Emory. New York: Dover.

Lockwood, Michael. 2005. *The Labyrinth of Time*. Oxford: Oxford University Press.

Lucas, John. 2008. "The Special Theory and Absolute Simultaneity." In *Einstein, Relativity and Absolute Simultaneity*, edited by Quentin Smith and William Lane Craig, 279–290. London: Routledge.

Maudlin, Tim. 1996. "Spacetime in the Quantum World." In *Bohmian Mechanics and Quantum Theory: An Appraisal*, edited by J.T. Cushing, Arthur Fine, and S. Goldstein, 295–305. Boston: Kluwer Academic.

Maudlin, Tim. 2002. *Quantum Non-Locality and Relativity*. 2nd ed. Malden: Wiley-Blackwell.

Maudlin, Tim. 2008. "Non-Local Correlations in Quantum Theory: How the Trick Might Be Done." In *Einstein, Relativity and Absolute Simultaneity*, edited by William Lane Craig and Quentin Smith, 156–179. London: Routledge.

Merali, Zeeya. 2009. "Splitting Time from Space—New Quantum Theory Topples Einstein's Spacetime." *Scientific American*, December, 18–21.

Misner, Charles W., Kip S. Thorne, and John Archibald Wheeler. 1973. *Gravitation*. San Francisco: W.H. Freeman.

Monton, Bradley. 2006. "Presentism and Quantum Gravity." In *The Ontology of Spacetime*, edited by Dennis Dieks, 263–280. Boston: Elsevier.

Monton, Bradley. 2011. "Prolegomena to Any Future Physics-Based Metaphysics." In *Oxford Studies in Philosophy of Religion Volume 3*, edited by Jonathan L. Kvanvig, 142–165. Oxford: Oxford University Press.

Ohanian, Hans C. 2007. "The Real World and Space-Time." In *Relativity and the Dimensionality of the Word*, edited by Vesselin Petkov, 81–100. Montreal: Springer.

Penrose, Roger. 1989. *The Emperor's New Mind*. Oxford: Oxford University Press.

Petkov, Vesselin. 2009. *Relativity and the Nature of Spacetime*. 2nd ed. Berlin: Springer.

Pitts, J. Brian. 2004. "Some Thoughts on Relativity and the Flow of Time: Einstein's Equations given Absolute Simultaneity." *Preprint*. http://philsci-archive.pitt.edu/2760/. Accessed January 18, 2009.

Putnam, Hilary. 1967. "Time and Physical Geometry." *Journal of Philosophy* 64: 240–247.

Rickles, Dean. 2008. "Quantum Gravity: A Primer for Philosophers." In *The Ashgate Companion to Contemporary Philosophy of Physics*, edited by Dean Rickles, 262–382. Aldershot: Ashgate.

Sklar, Lawrence. 1981. "Time, Reality and Relativity." In *Reduction, Time, and Reality*, edited by Richard Healey, 129–142. Cambridge: Cambridge University Press.

Smith, Sheldon R. 2000. "Resolving Russell's Anti-Realism about Causation." *The Monist* 83 (2): 274–295.

Smolin, Lee. 2001. *Three Roads to Quantum Gravity*. New York: Basic Books.

Speziali, Pierre, editor. 1979. *Albert Einstein, Correspondance Avec Michele Besso, 1903–1905*. Paris: Hermann.

Stein, Howard. 1968. "On Einstein-Minkowski Space-Time." *Journal of Philosophy* 65 (1): 5–23.

Timpe, Kevin. 2008. *Free Will: Sourcehood and Its Alternatives*. London: Continuum.

Torretti, Roberto. 1996. *Relativity and Geometry*. New York: Dover Publications.

Weinberg, Steven. 1972. *Gravitation and Cosmology: Principles and Applications of the General Theory of Relativity*. New York: John Wiley & Sons, Inc.

Zagzebski, Linda. 2000. "Does Libertarian Freedom Require Alternate Possibilities." *Philosophical Perspectives* 14 (s14): 231–248.

Zimmerman, Dean. 2011. "Presentism and the Space-Time Manifold." In *The Oxford Handbook of Philosophy of Time*, edited by Craig Callender, 163–244. New York: Oxford University Press.

4

Divine Action and the
Laws of Nature

4.1   Divine Intervention(?)

Magician David Blaine bothers me. For others, it's Criss Angel. I know they can't actually levitate, walk on water, twist their hands in 360° circles, or throw cards through glass, but it certainly *looks* that way. So how do I know it's all just an illusion? Because if their tricks were real, they would violate one or more laws of physics and that can't happen. Among other things, the physical sciences lay out the boundaries of what is possible and what is not.[1]

Traditional theists do not believe that this same argument applies to God. In the Hebrew Bible, for example, one finds axe heads that float (2 Kings 6) and rocks producing water (Num 20) presumably because God caused these events. People cannot violate the laws of nature, but most theists believe that God can. God sometimes overrides the natural capacities of objects in order to produce an outcome that would not have happened naturally.

This traditional view—what some would call the vulgar or naive view—has been under fire for some time now and not just by atheists. Theologians and philosophers have argued against an interventionist view of divine action for centuries. Although God could intervene in the natural order, they believe that God does not and will not. Intervention would only be needed if things weren't going the way God wanted, but how could nature *not* behave according to the will of its omnipotent creator? Consider the philosopher G.W. Leibniz writing in the early 1700s:

The Physics of Theism: God, Physics, and the Philosophy of Science,
First Edition. Jeffrey Koperski.
© 2015 John Wiley & Sons, Ltd. Published 2015 by John Wiley & Sons, Ltd.

If active force should diminish in the universe by the natural laws which God has established, so that there should be need for him to give a new impression in order to restore that force, like an artist's mending the imperfections of his machine, the disorder would not only be with respect to us but also with respect to God himself. He might have prevented it. (Leibniz and Clarke 1956, 29)

The idea is that God would not create a world that had to be wound up occasionally, like a watch. The need for maintenance would be an "imperfection" not only in creation but also in its creator. If things weren't going the way God wanted, instead of having to reach in and change things, God could have "taken better measures to avoid such an inconvenience, and therefore, indeed, he has actually done it" (1956, 29). An omniscient, omnipotent creator would have set things up in the beginning to bring about the desired events. For example, if God wanted a bright light to shine in the sky during Jesus' birth, Leibniz's God would not wait until the first century to supernaturally create one. God would simply have set things up so that, say, a supernova would occur at just the right place and time to produce the desired phenomenon.

By the twentieth century, most theologians had come to agree that Leibniz was at least partially right: God would not violate his own laws of nature. At the same time, this view seemed incomplete. The creator/sustainer-but-nothing-else view seemed a bit too passive, too deistic. The search began for a view of divine action that allows for active, ongoing governance without violating the laws of nature.

There are now many noninterventionist views of special divine action. Such acts are "special" in the sense that they are something other than God's continual sustaining of the universe. The views are "noninterventionist" in the sense that they show how God can influence the natural order without violating the laws of nature.

So where is the physics in all this? It shows up in the guise of determinism: Is there any real contingency in nature or not? If determinism is true, then each physical event that occurs *must* occur given the circumstances and events leading up to it. The problem is that a deterministic universe does not seem to leave any room in which God might act. If the outcome of every event is already fixed by the laws of nature, then God can't act within nature without violating those laws. Noninterventionists look for areas where determinism does not rule, which, as you might have guessed, will take us into quantum mechanics.

While all this might seem rather intuitive, the devil, as they say, is in the details. (Depending on one's perspective, the metaphor here is either somewhat unfortunate or entirely apt.) In fact, the very notion of "violating the laws of nature" is harder to pin down than one might imagine. For example, say we define an intervention as any event not determined by the laws of nature. Most theists would say the event = 'the coffee mug steams on my desk' would not have happened without God's continual sustaining of the universe. Hence, sustaining the universe would count as an intervention, which is not what noninterventionists had in mind (Plantinga 2008, 388) (more on that problem later). First, let's consider the arguments against the traditional, interventionist view of divine action (Section 4.2). That will be followed by a brief discussion of the laws of nature (Section 4.3), some specific noninterventionist models (Section 4.4), and an assessment of the most popular such model based on quantum mechanics (Section 4.5). The chapter ends with a reexamination of the motives for noninterventionism (Section 4.6) and a new way to understand what the fight is about (Section 4.7).

4.2   The Problems with Intervention

There are five main reasons why divine intervention has come under fire in recent decades.

4.2.1   An Incompetent God

The first argument is rooted in the Leibnizian clockmaker analogy touched on earlier. Any divinity that occasionally needed to fix the way nature works would seem to be less than omniscient or omnipotent. After all, couldn't God have gotten it right in the first place? Why the need to fix things after the fact? Arguing against those who think God violate the laws of nature on occasion, Leibniz says,

> According to their doctrine, God Almighty wants to wind up his watch from
> time to time: otherwise it would cease to move. He had not, it seems, sufficient
> foresight to make it a perpetual motion. Nay, the machine of God's making is
> so imperfect, according to these gentlemen; that he is obliged to clean it now
> and then by an extraordinary concourse, and even to mend it, as a clock-
> maker mends his work; who must consequently be so much the more

unskillful a workman, as he is oftener obliged to mend his work and to set it right. (Leibniz and Clarke 1956, 11–12)

Those who hold such a view "have a very mean notion of the wisdom and power of God." More recently, William Pollard has argued that intervention would be

> … like seeing a great actor stop in the midst of a magnificent performance to pick up a line from a prompter, or a master craftsman tampering awkwardly with an otherwise perfect creation. Anyone who has had the privilege of having the whole marvelous structure of mathematical physics unfolded before his imagination and experienced the thrill of it cannot fail but find the thought of such intervention shocking. (Pollard 1958, 28–29, quoted in Saunders 2000, 530)

Again, a fully competent creator would not need to intervene.

If a miracle is, as David Hume defined it, a violation of the laws of nature, then many noninterventionists prefer marvels to miracles. Marvels are amazing and improbable events that indicate God's favor without breaking any natural laws. A marvel is what philosopher Thomas Tracy describes as a "subjectively special act of God" (2008, 603) or as Pollard says,

> Biblical miracles are, like that in the exodus, the result of an extraordinary and extremely improbable combination of chance and accident. They do not, on close analysis, involve, as is so frequently supposed, a violation of the laws of nature. (Pollard 1958, 115, quoted in Saunders 2000, 531)

For example, consider an alternative history where first-century Jews in Jerusalem fasted and prayed for 40 days for God to spare them from the invasion of the Roman general Titus. Say then that 1 day later Titus was struck by a meteorite and died. A miracle? Not necessarily. God might have foreseen these prayers and set this meteor on its path at creation so that it would land on Titus' head at just the right moment—a marvel, not a miracle. If one is creative enough, most and perhaps all miracle claims can be naturalized into marvels, as philosopher Michael Ruse shows:

> Many if not all of the miracles happened according to law; their miraculous nature comes from their meaning or significance. The every-day miracles of the Gospels—turning the water into wine and feeding the five thousand and even raising Lazarus—can be explained as the enthusiasm of the moment.

People's hearts were so filled with love by Jesus' talk and presence that spontaneously and out of character they shared their food. To think otherwise—to think that Jesus actually turned loaves and fishes into a banquet—is if anything a bit degrading, making the Redeemer a kind of high-class caterer. Lazarus and the ruler's daughter were more likely brought back from trances. They may have been dead to all intents and purposes, and Jesus' actions were highly significant, but one should not suppose that Lazarus and the girl were necessarily clinically dead. (Ruse 2001, 96)

Most noninterventionists aren't willing to go quite this far. They want an avenue for special divine action—by definition acts that go beyond the creation and sustaining of the universe—but one that does not entail a violation of natural law.

4.2.2   *A Capricious or Inconsistent God*

Another worry is that the theistic God isn't supposed to be capricious like the Greek or Norse gods: sometimes they act, sometimes not, based merely on whim. The theistic God, in contrast, is supposed to be trustworthy, not capricious, and this trustworthiness is reflected in the laws of nature themselves as John Polkinghorne argues:

Their discerned regularities are pale reflections of his faithfulness towards his creation; they are expressions of his acquiescent and economic wills. He will not interfere in their operation in a fitful or capricious way, for that would be for the Eternally Reliable to turn himself into an occasional conjurer. (Polkinghorne 1989, 24)

When God acts, he does so for good reasons and in a manner that reflects his trustworthiness. By analogy, we trust our fire department because they always respond when called. If firefighters only responded when they felt like it, that would not engender a sense of security. Any model of divine action must capture the consistency and reliability of a good and omnipotent God, says Polkinghorne:

Two general conditions must surely apply to any adequate account of divine action. The first is that it must be continuous and not fitful, correctly referred to as "interaction" rather than "intervention." There can be nothing capricious or occasional in God's activity. (Polkinghorne 1996, 244)

Polkinghorne demands that models of divine action fit naturally with God's sustaining of the universe—a continuous, ongoing act. God does not infuse or recharge nature with the capacity to exist every so often. However it might seem to us, God is always interacting with nature. Divine action is constant, not episodic, more like changing one's rate of breathing than setting off a firecracker.

A closely related worry is that violating the laws of nature would mean that God is inconsistent. On one hand, God has declared the laws of nature, endowing them with nomological necessity. On the other hand, God decides to violate the very laws he has ordained. Miraculous divine action thus entails a conflict within the divine will, noninterventionists argue, a view that goes back at least as far as Spinoza (1632–1677):

> Now, as nothing is necessarily true save only by, Divine decree, it is plain that the universal laws of nature are decrees of God following from the necessity and perfection of the Divine nature. Hence, any event happening in nature which contravened nature's universal laws, would necessarily also contravene the Divine decree, nature, and understanding; or if anyone asserted that God acts in contravention to the laws of nature, he, ipso facto, would be compelled to assert that God acted against His own nature—an evident absurdity. (*Theologico-Political Treatise* part 2, chap. 6)

This argument has not been left to history. A similar point is made by physicist Robert J. Russell:

> The problem with interventionism is that it suggests that God is normally absent from the web of natural processes, acting only in the gaps that God causes. Furthermore, since God's intervention breaks the very processes of nature which God created and constantly maintains, it pits God's special acts against God's regular action, which underlies and ultimately causes nature's regularities. (2008a, 584)

Inconsistency on the part of parents or perhaps the legal system is just part of our finiteness and fallibility. The conclusion here is that no such inconsistency should be found within the will of an omniscient being. Whatever one's model of divine action, so say noninterventionists, it must maintain theological consistency, which the traditional view of miracles does not.[2]

4.2.3   *The Problem of Evil*

The next issue is well known and overlaps with the capriciousness objection: How can an omnipotent, omniscient, and—most importantly—omnibenevolent God allow so much evil in the world? While there are many ways to attack this problem, noninterventionists argue that a naive view of divine action makes things worse:

> The problem of allowing miraculous intervention to turn water into wine, to heal the sick, to raise the dead, or to alter the weather is that this involves either a suspension or alteration of the natural order. Thus, the question arises as to why this happens so seldom. If this is allowed at all to achieve some good, why is it not allowed all the time, to assuage my toothache as well as the evils of Auschwitz? (Ellis 1995, 383)

Believing that God sometimes miraculously intervenes to prevent evil, Ellis argues, produces a slippery slope. On what basis does God occasionally act and then not help others in the same circumstances? I have a former student who believes that God healed him of a brain tumor, and the sequence of events is certainly interesting. With surgery scheduled, he went to a prayer service held on his behalf for the healing of the tumor. Later that week, a final scan showed that the tumor was gone. What's wrong with my student's belief that God miraculously removed the tumor? The first thing to note is that it would have required no more effort on the part of an omnipotent being to also stop the Asian tsunami of 2004. Why would God heal one tumor but not save the lives of hundreds of thousands of people? Or more personally, why did God heal my student's brain tumor but not my brother's heart disease?

Polkinghorne's answer is similar to the free will defense for moral evil.[3] When it comes to natural evil like disease and earthquakes, there is a "free-process defense":

> God allows the whole universe to be itself. Each created entity is allowed to behave in accordance with its nature, including the due regularities which may be part of that nature. [Quoting a previous publication], "God no more expressly wills the growth of a cancer than he expressly wills the act of a murderer, but he allows both to happen. He is not the puppet master of either men or matter." (Polkinghorne 1994, 83)

If God were to restrain a thief from snatching your laptop, this would prevent an evil act, but it would also override the thief's free will. Many theists believe that this freedom has such a high value that God permits evil decisions. The suggestion here is that God likewise allows nature itself to have a kind of freedom and autonomy. The laws of nature are such that mountains rise under plate tectonics, rain falls as part of the water cycle, and rivers carve spectacular valleys via slow erosion. But these same laws produce volcanoes, hurricanes, and floods, all of which sometimes kill people.

Noninterventionists argue that there is a reason why God doesn't intervene in these cases and this in turn provides a reply to the problem of evil for natural disasters and disease. God doesn't intervene because doing so would violate the freedom of nature itself. The argument requires that, like the free will granted to humans, the freedom of nature to act according to its God-endowed capacities has intrinsic value. While one might question the truth of that premise, consider the alternative: if one allows for miraculous divine action in some cases, then there is no answer to the problem of natural evil. Philosophical theology thus again weighs in favor of noninterventionism.

4.2.4   *The God of the Gaps*

I once overheard a seminary student argue that "we still don't know how atoms hold together. Since like charges repel, all those protons packed together in a nucleus should blow apart." One can only conclude, he thought, that God supernaturally holds atomic nuclei together. This (somewhat unfortunate) argument fits a pattern: we observe x, but x is unlikely or impossible based on current scientific understanding; hence, God is at least partially responsible for x. The phrase 'God of the gaps' refers to using God to explain the gaps in our knowledge in this way. The term is almost always used pejoratively—a rhetorical tactic to be avoided. Why? Well, consider the student's argument about atomic nuclei. The problem was that his science was several decades out of date. We've long known that the strong nuclear force holds atomic nuclei together, overwhelming the repulsion of like charges.

This, then, is the worry about God-of-the-gaps explanations. If we explain a gap in our knowledge by appealing to God, and then further research fills that gap with a naturalistic cause, we no longer need God. The short-term apologetic benefits are outweighed by the long-term undermining of theism. The poster child for this is Isaac Newton, who believed that the gravitational pull of the planets on one another should destabilize their orbits.[4] How is it that the planets have maintained their orbits over the aeons? In addition, why have the stars not fallen into one another over time? Their mutual attraction and irregular spacing should cause collisions.[5] The answer is that God is responsible:

> [Does] it not appear from Phænomena that there is a Being incorporeal, living, intelligent, omnipresent, who in infinite Space, as it were in his Sensory, sees the things themselves intimately, and throughly perceives them, and comprehends them wholly by their immediate presence to himself. ... And though every true Step made in this Philosophy brings us not immediately to the Knowledge of the first Cause, yet it brings us nearer to it, and on that account is to be highly valued. (*Opticks*, Query 28)

Unfortunately, for Newton, these gaps in our knowledge were eventually closed, although it took about 200 years to do so.[6] Therein lies the problem. When science eventually closes these epistemic gaps, God gets pushed out. Those opposed to God-of-the-gaps reasoning believe that such arguments are harmful to theism. If one relies on such gaps as justification for belief in God, as the seminary student was doing, and then those gaps are closed, theistic belief is undermined. Hence, it is far better scientifically and theologically not to employ such reasoning.

All this is relevant here since miraculous divine action seems to invite gap arguments. If one believes that God brings about miracles, as opposed to mere marvels, then natural causes are by definition not sufficient to explain those events. The temptation, then, is to posit God as the explanation for events that are merely puzzlingly and that do not immediately appear to have a scientific explanation. When a naturalistic cause is eventually discovered and the gap is closed in time, God is pushed out. Noninterventionists believe that this is losing strategy for theism. In academic circles, models of divine action with any hint of gap reasoning are often dismissed out of hand.

4.2.5   Conflicts with Science

Finally, there are two ways in which interventionism seems to clash with modern science.

The first is a conflict with the metatheoretic shaping principle (Section 1.3.1) known as *methodological naturalism*: science can only appeal to natural laws and physical entities as explanations of observable phenomena.

Scientists must therefore not look beyond their own domains in the search for explanations. Note that methodological naturalism is not the same as metaphysical naturalism, the view that the only entities, systems, and causes are natural ones.

Methodological naturalism, many claim, is in conflict with interventionist divine action. Good science cannot appeal to supernatural causes. Hence, any reference to miracles is unscientific. Alvin Plantinga (2011, 71) cites theologian John Macquarrie as an example of this line of thought:

> The way of understanding … supernatural interventions belongs to the mythological outlook and cannot commend itself in a post-mythological climate of thought …. The traditional conception of miracle is irreconcilable with our modern understanding of both science and history. Science proceeds on the assumption that whatever events occur in the world can be accounted for in terms of other events that also belong within the world; and if on some occasions we are unable to give a complete account of some happening … the scientific conviction is that further research will bring to light further factors in the situation, but factors that will turn out to be just as immanent and this-worldly as those already known. (Macquarrie 1977, 248)

Macquarrie is far from unique. Thomas Tracy argues that for many theologians, the deduction is straightforward:

1. The idea of miraculous divine intervention is no longer acceptable to modern human beings whose understanding of the world has been shaped by the sciences.
2. Science commits us to understanding events as occurring within an unbroken continuum of natural causes. [methodological naturalism]
3. Any act of God that alters the course of events in the world will disrupt the causal continuum of nature: i.e. it will be a miraculous intervention.
4. So the idea of particular divine action in the world must be given up. (Tracy 2008, 599)

Miracles are thus unscientific in the strong sense. It is not merely that science gives us no reason to believe in them. Miraculous intervention is forbidden by modern science.

The second conflict is with a specific law, not a shaping principle. As physicist William Stoeger argues, miraculous divine action violates the conservation of energy. The laws of nature, he says, do

not easily allow for divine intervention—at least not direct divine intervention—because that would involve an immaterial agent acting on or within a
material context as a cause or a relationship like other material causes and
relationships. This is not possible; if it were, ... energy and information
would be added to a system spontaneously and mysteriously, contravening
the conservation of energy ... (Stoeger 1995, 244)

Since conservation of energy is as well established as any law in physics,
the conflict here is stark. For God to directly influence a physical system,
energy must be introduced into that system—a clear violation of
established science. Noninterventionists take such conflicts seriously. It's
one thing for theology to have tensions with science at an abstract metaphysical level. It is quite another when theology allows for acts forbidden
by physical law. Any viable model of divine action should mesh with
science, not contravene it.

4.3   The Nature of the Laws of Nature

We saw earlier how the idea that nature has laws was rooted in theism
(Section 1.2.2). While most scientists today do not make a connection between natural laws and God,[7] the nature of those laws has long been a controversy in philosophy. Some distinctions are needed in order to understand
the noninterventionist program.

4.3.1   *Regularity versus Necessitarian Laws*

The standard *regularity* accounts hold that the laws merely summarize the
uniformities we observe in nature. Most regularity theorists believe that the
objects identified in science—electrons, genes, etc.—exist and that those
entities tend to behave in predictable ways. Metaphysically, there is nothing
more to the laws than the regular behavior of those entities. In particular,
laws do not literally "govern" the way objects behave on the regularity view;
the laws are just the uniform ways the things in nature in fact behave. The
laws themselves are determined by entities and their behavior. If there were
no physical beings, there would be no laws.

A different regularity view is that laws are those generalizations that are
most fundamental in our scientific thinking (Earman 1986). If we tried to
organize all scientific knowledge from the most specific claims (small

scope) to very general claims (wide scope), the laws would be those generalizations in which everything else is a special case, something like the axioms in Euclidean geometry. The laws of nature are those truths that cover and systematize as much scientific knowledge as possible. This view takes lawlikeness to be a matter of our theorizing, rather than the constitution of nature itself. We invent the laws as we organize our knowledge. The laws aren't "out there" to be discovered.

Necessitarian accounts, on the contrary, take it that nature is in some sense governed by its laws. The laws on this view have their own metaphysical standing; they do not depend on the behavior of entities. Even in an empty universe, for example, it would still be a law that opposite charges attract each other. Necessitarian laws thus have a kind of abstract existence, something akin to Plato's Forms (see Section 1.2.4). Lawlikeness on this view is part of the foundation of nature itself and in no way depends on our theorizing or the prior existence of concrete entities.

I should mention that some philosophers of science think that the very idea of laws of nature is overblown. Bas van Fraassen has argued against the notion on empiricist grounds (1989). Others, like Nancy Cartwright, believe that the "governing" aspect ascribed to laws is actually carried out by other things such as "capacities" and "natures" (1989). For our purposes, these alternative views can be ignored. The point is merely that many of the discussions about divine intervention fail to distinguish which interpretation of natural law is in play.

4.3.2 *Laws and Law Statements*

Another distinction is between the laws themselves, out there in nature, and what we believe them to be. Start with the former. If laws are understood in a necessitarian way, then there is a fact of the matter about what they are. The laws are a part of nature—things to be discovered. Consider Boyle's law from elementary chemistry: $PV = k$. In an ideal gas at a constant temperature, the pressure times the volume will remain constant. Notice that scientists could have been either right or wrong about Boyle's law. Either gases really do behave this way or not. The expression '$PV = k$' is not itself a law of nature. The equation that appears in the textbooks is a *law statement*. It is the expression of what we believe one law of nature to be.

The distinction between law and law statement is really quite simple. It is roughly the same as between a belief someone might have and a fact of the matter in reality. Beliefs can be true or false; reality is just the way it is. So while I might believe that there is groundhog in my backyard, there either is or there isn't. The belief might be true or false, but there is no sense in which the groundhog can be true or false. It's either out there, digging up my wife's garden, or it isn't.

Why is this important? First, it's a matter of clarity. There is a lot of hand-wringing in the divine action literature about God "breaking the laws of nature." What precisely does that mean? Second, there is a danger of equivocation. Some arguments might sound stronger than they should because the meaning of 'law' is subtly sliding from one category to another. So let's be clear then. When noninterventionists worry about violating "the laws of nature," they must mean *laws* rather than *law statements*. If God were to "break" a mere law statement, that would just mean that the act was contrary to what we believe to be a law of nature. Perhaps we got it wrong. Hence, noninterventionists should be concerned with violations of the actual laws of nature, regardless of whether those laws correspond to the law statements in the textbooks. However, this immediately presents a problem. As progress in all of the sciences shows, we don't know the laws of nature in a completely accurate and precise way. How then would we know that an actual law was violated? For any seemingly miraculous event, it could be the case that a law of nature was at work that we are not familiar with. Hence, the noninterventionist claim must be that God never violates the actual laws of nature, whatever they happen to be. Apparent violations of textbook law statements merely indicate a lack of knowledge on our part.

Violation talk also only makes sense under the necessitarian view of laws. On the regularity account, the laws describe how entities behave, but the laws themselves have no independent reality of their own. In the philosophical jargon, regularity laws *supervene* on entities and events; laws are only real in a derivative sense. If the universe were empty, there would be no natural laws under the regularity view. There is no sense in which God might suspend the laws of nature taken as regularities, since the laws themselves are nothing more than the way things normally behave. Strictly speaking, regularity-based laws don't exist in such a way that they could be broken. There literally isn't anything "out there" to be broken.[8]

In sum, when noninterventionists talk about violations of natural laws, their concerns only make sense if we interpret these as necessitarian laws, rather than law statements or laws understood as regularities.

As we've seen, noninterventionists have a wide range of philosophical, theological, and scientific concerns. Acceptable models of divine action, in their view, must avoid conflicts with science and any hint of gap reasoning without implying that God is capricious or inconsistent. Let's now consider some concrete proposals.

4.4   Noninterventionist Divine Action

As the reader might have guessed, this is the section where quantum mechanics comes into play. To see why, we first need to consider some of the limitations of classical mechanics.

4.4.1   Avoiding Determinism

Newtonian physics, we are often told, is deterministic: every event is governed by the laws of nature without exception:

> In classical physics, the fundamental laws were deterministic and implied, philosophically, that nature itself is deterministic, a closed causal system of forces rigidly determining the motion of matter. (Russell 2008a, 580)

While many events look random from our limited perspective, in classical physics, there is no physical contingency. Every physical event is nomologically necessary: it must have the outcome that it does have given the prior state of the system and the laws of nature.

Consider dice in a Newtonian world. We say that the way dice tumble when rolled is a random event, but notice that this sense of randomness is merely a matter of ignorance on our part. If we knew the angular and linear momentum imparted to the dice when tossed, the force of gravity, and the coefficient of friction of the table, we could say precisely how the dice will stop. Given the laws of nature, they must stop precisely as they do. There is no randomness here in terms of the physics. The throwing of the dice is a fully deterministic event. In that sort of world, what we call a "random event" is merely a reflection of our ignorance.

The dice illustration is just the beginning. Consider a "Laplacian demon" (after physicist Pierre Laplace): an idealized intelligence with unlimited computational capacity that has full knowledge of the laws of nature and the state of the universe at a specific time. In a deterministic, Newtonian world, such a being could predict the state of the entire universe arbitrarily far into the future. Since no event escapes the laws of nature and Laplacian demons do not suffer from our lack of information, they would have no trouble solving the relevant equations and thereby knowing the outcome of any future event.

Noninterventionists argue that determinism would put severe constraints on what God could do in nature. If the laws of physics already provide sufficient conditions for each event, there isn't any room left in which God might act. Philosopher Nicholas Saunders argues that in

> a totally deterministic world … the causal nexus of science is drawn so tight that there is no real freedom for either God or human beings. In such a world Laplace's famous demon reigns supreme. God cannot act in any creative way through the causality of science and still remain true to the deterministic rules put in place at creation. (Saunders 2000, 254)

In other words, any act of God within a deterministic universe must violate the very laws that God implemented at creation.

The upshot of all this is that noninterventionists must find some degree of *indeterminism* in nature. For God to act, it cannot be that every physical event has a sufficient physical cause—a premise that all prominent noninterventionists agree on, as Saunders (2000) shows. Consider philosopher–theologian Keith Ward:

> [If] God acts (brings changes about intentionally) [then] there are states of the physical universe which are not sufficiently explained by the operation of physical causes alone. To put it in the words of the crude formulation, there must be gaps in physical causality, if God is ever to do anything. (1990, 77)

Note that these are not the explanatory gaps we considered in the God-of-the-gaps problem. Ward's gaps are not due to a lack of knowledge on our part. He is talking about a real, metaphysical openness such that not every physical event has a sufficient cause. An indeterministic universe has a kind of intrinsic, irreducible randomness among physical events. Even a Laplacian demon would only be able to calculate the probability of such events; it could not predict their outcome with certainty. Where might one find this sort of deep, metaphysical indeterminism? Quantum mechanics to the rescue.

4.4.2   Quantum Indeterminism

While there are many interpretations of quantum mechanics, the orthodox view provides just the sort of intrinsic randomness needed. On this view, some quantum events do not have a sufficient cause. The probabilities one encounters in quantum mechanics are metaphysical, not merely epistemic; they are real, irreducible aspects of nature itself, not merely a reflection of our ignorance. Not even a Laplacian demon can know when a given atom of uranium will decay, since that event is not fully determined by prior conditions and the laws of physics.

Many noninterventionists look to take advantage of this quantum indeterminism. Say we have pair of quantum dice: two 6-sided dice that roll in a metaphysically random way, just as I've described quantum mechanics. On each roll, there are 36 physically possible outcomes. Since each outcome is possible, there can be no violation of any law no matter which number is rolled. Now, what if God wanted the dice to roll a 12? In that case, God could influence the system in order to get the desired result without breaking any laws. So long as God works within the range of possibilities presented by quantum mechanics, there is no need to override the laws of nature. The laws and initial conditions do not determine a precise outcome. Since rolling a 12 is a physically possible result, God merely brings about this one outcome from among the indeterministic possibilities.

On the orthodox interpretation, there are many quantum events analogous to the dice. A wide range of random effects is physically possible. God is therefore able to influence their outcome. Of course, these random events occur on a very small scale, which is why our everyday experience seems perfectly Newtonian. Nonetheless, as the number of small events adds up, they eventually produce observable results. Hence, God can govern the physical world through quantum indeterminism without breaking any of the laws God has put into place. Let's call this view of divine action *quantum determination* (QD).

There are many examples of noninterventionists appealing to quantum mechanics in this way, going back at least as far as Pollard's (1958). Let's consider two, physicist Robert J. Russell and philosopher Nancey Murphy:

[The] presence of statistics in these fields arises not from our ignorance of the underlying deterministic forces but from the fact that there are, in reality, no sufficient underlying forces or causes that fully determine particular physical processes, events, or outcomes. (Russell 2008a, 581)

If I adopt the interpretation that these quantum statistics reflect ontological indeterminism, then I may argue that God can act together with nature to bring about all events at the quantum level, and that these events give rise to the classical world. (Russell 2002)

> God's governance at the quantum level consists in activating or actualizing one or other of the quantum entity's innate powers at particular instants. (Murphy 1995, 342)

In fact, Murphy believes that "God is involved in every random quantum event" (Murphy 1995, 339). How so? In the sense that the everyday objects we experience are composed of quantum mechanical constituents:

> Macro-level objects are complex organizations of their most basic constituents (this is analytic). To a great extent, the behavior of the whole is determined by the behavior of its parts. So the laws that describe the behavior of the macro-level entities are consequences of the regularities at the lowest level …. (Murphy 1995, 346)

> So our theological intuitions urge upon us the view that, in *some* way, God must be a participant in *every* (macro-level) event. God is not one possible cause among the variety of natural causes; God's action is a necessary but not sufficient condition for every (post-creation) event. In addition, I claim that God's participation in each event is *by means of* his governance of the quantum events that *constitute* each macro-level event. (Murphy 1995, 343)

Murphy and Russell hold that God both sustains the existence of the quantum world—and thereby everything else—and influences every random quantum event. They argue that this model of divine action is therefore not episodic, but continuous. God is always acting, rather than reaching into creation every so often. If God wants to change the trajectory of natural causes without violating natural law, quantum mechanics provides just the kind of metaphysical indeterminism needed.

Twentieth-century physics was kind to noninterventionists in more than one way. It not only provided quantum mechanics but also a later discovery that would help harness the power of QD. A surge of research in the 1980s produced what some call the third great theory of the century: chaos theory.[9]

4.4.3   Quantum Mechanics + Chaos

For reasons to be discussed in Section 4.5.4, noninterventionists have tried to find ways of leveraging quantum events. One reason is that the randomness in quantum mechanics is naturally quite small and hidden from view. Hence, if God is going to do any significant governance by means of quantum mechanics, somehow those events are going to have to be amplified. Enter chaos theory.

The doctrine that small causes bring about small effects is grounded in our experience of linear systems. Turning up the volume knob on a radio increases the power output of the speakers proportionally. This simple input–output rule does not apply to nonlinear chaotic systems, which can behave in complex, unpredictable ways. The central notion behind chaos is *sensitive dependence on initial conditions*, a.k.a. "the butterfly effect." The idea is that if the atmosphere evolves chaotically, then a butterfly flapping its wings in Japan today might be sufficient to change the weather in Miami sometime next year from what would have been a sunny day into a hurricane. In other words, extremely small causes can have large-scale effects. Any slight perturbation in a chaotic system will produce a dramatic change in the future state of that system. How slight? When astronomers plot the orbits of the planets, they ignore the gravitational pull of everyday objects. Mount Rushmore, my house, and Maggie (my in-law's yappy dog) are too small to make a difference. If, however, Maggie suddenly disappeared from the surface of the Earth (thereby making it a more peaceful place), the gravitational change would significantly affect the motion of Hyperion, the chaotically tumbling moon of Saturn.

So how does this help the noninterventionist? Let's say that God does influence nature by determining the outcome of random quantum events. In a chaotic system, God need not affect innumerable quantum events to bring about observable effects. Chaos allows one QD to produce macroscopic results by changing the state of a nonlinear system. Under chaotic QD, God need not answer prayers for rain in Texas by miraculously creating a storm or even by manipulating the collapse of many trillions of quantum events. If global weather patterns are chaotic, then God need only make a particular quantum event fall one way rather than another and eventually this act will manifest itself as rain in Austin.

For those, like Murphy (1995), who endorse chaotic QD, it helps that chaos is ubiquitous in nature: weather systems, heartbeats, dripping faucets, predator–prey relations, tumbling asteroids, etc. In principle, God could take advantage of quantum indeterminacy and the butterfly effect to bring about the desired state of any chaotic system, all without violating the laws of physics.

4.4.4   Related Views

There are other noninterventionist proposals, although none rivals the popularity of QD. Before moving on, let's briefly consider three of these.

4.4.4.1   John Polkinghorne's Open Universe

One of the most influential figures mentioned so far is John Polkinghorne, a former particle physicist who is now an Anglican priest. Polkinghorne is a friendly critic of QD. He agrees with the idea that God's violating of the laws of nature reduces divine action to a mere "alternative source of energetic causation, competing with the effects of physical principles from time to time and overriding them" (Polkinghorne 1989, 34). On the other hand, Polkinghorne does not believe that divine action is limited to discrete events, as it seems to be in QD. God does not merely manipulate physical conditions to bring about some observable event in the future (Polkinghorne 1995, 152–154).

Instead, Polkinghorne takes the intrinsic unpredictability of nature dictated by chaos and quantum mechanics as pointing to something deeper. Physics has shown that there are strict limits to predictability, a limitation on our knowledge. As we've seen, not even a Laplacian demon can know the outcome of some quantum events; chaos only makes things worse. This limit on knowledge, Polkinghorne argues, has metaphysical implications, namely, an additional ontological openness that goes beyond current physics:

> If you are a realist and believe, as I believe, that what we know (epistemology) and what is the case (ontology) are closely linked to each other, it is natural to go on to interpret this state of affairs [i.e., the unpredictability of chaos and quantum mechanics] as reflecting an intrinsic openness in the behaviour of these systems. (Polkinghorne 1989, 29)

On Polkinghorne's view, the proven limits on physical knowledge are indicators of still hidden degrees of freedom in nature. Chaos and quantum mechanics point to undiscovered ways in which God can influence the universe without intervention.

The respect many have for Polkinghorne has not translated into followers who are anxious to take up his banner and run with it. While his "physicist's intuition" tells him that metaphysical implications should be drawn from the limits on knowledge that nature has imposed, there aren't many physicists or philosophers following suit. Most take Polkinghorne's view as more of an interesting idea than a concrete proposal—gesturing toward the right answer without saying precisely what that answer might look like.

4.4.4.2   Deck Stacking

Philosopher Michael Murray has argued that instead of intervening in nature, an omniscient creator could foresee how to set things up in advance.

Like stacking a deck of cards to favor one player, God could create the world in such a state that it would evolve into a desired outcome (2003, 315–323). Even in a world with quantum indeterminacy, God could initiate several causal chains with a high probability of producing a planet like ours in due time.

4.4.4.3   *Emergence*

The different sciences study nature at different scales or levels: biology, from ecosystems down to cells; chemistry, from the chemical constituents of cells down to atoms; and physics, from atoms down to the most fundamental level (if there is one). A recent trend is to see these levels as more than a convenient division of labor among the sciences. Some take the levels to be metaphysically significant, interacting with each other in complex ways. High-level macroscopic phenomena, such as the directionality of entropy (i.e., the second law of thermodynamics), emerge from the level of atomic physics. Some also believe that the higher levels influence the lower in a top-down fashion. For example, the atoms in my finger move the way they do not merely from local interactions with other atoms, but because they are part of an entire system. Events at the level of anatomy and neurophysiology causally influence events at the level of physics.

Similarly, Arthur Peacocke and others have argued that God might influence the universe in a top-down fashion, imparting information to the whole, rather than through the accumulation of small changes from the bottom (Peacocke 1993).

Current interest in top-down causation and emergence/reduction goes far beyond divine action. We leave it for now but will take up the topic again in more detail in Chapter 6. When it comes to divine action and physics, quantum mechanics and chaos play much more prominent roles in the literature. Although in my view, this high degree of enthusiasm is not warranted, we should examine some of the details first.

4.5   QD: Pro and Con

4.5.1   *What Quantum Mechanics Allows*

The good news for the QD model of divine action is that, under some interpretations of quantum mechanics, what it proposes is physically possible (more on which interpretations and why in Sections 4.5.2 and 4.5.3).

Even seemingly miraculous events can occur naturally on occasion. Let's start with a simple example. Say that we confine an electron within a hollow, highly charged metal sphere. Since like charges repel, a negative charge on the sphere will effectively force the electron to the center of the sphere and hold it there. Unless circumstances change, this particular electron is completely trapped. It will never gain enough kinetic energy to make it past the barrier.

One of the surprises of quantum mechanics is that the electron, contrary to classical electromagnetic theory, can "tunnel" through the charged sphere and reappear outside its shell. Instead of going over the mountain, as it were, the particle passes through it. Such an event isn't likely, but it can— and has—happened. Events that are impossible according to classical physics are possible under quantum mechanics. It is physically possible for the water molecules in my cup to simultaneously tunnel through the ceramic and land on my keyboard. (It's probably a good thing my youngest son doesn't know about quantum tunneling, lest he try this as an explanation.)

Now consider the purported healing of my student's brain tumor (Section 4.2.3). It is physically possible, although highly unlikely, that all of the particles making up his tumor would tunnel through his head and disperse into the atmosphere never to be seen again, just like the trapped electron. If such an event is physically possible and so by definition not a violation of natural law, God could bring about that event without overriding any laws of nature. From the vast space of possibilities allowed by quantum mechanics, God can select one with the net effect that the tumor is gone. Such an act counts as special divine action since God had a hand in the outcome, but it requires no intervention because the entire episode is physically possible. It's not hard to generalize this idea to cover a whole range of seemingly miraculous events, as philosopher of physics John Earman suggests,

> If we try to define a miracle as an event that is incompatible with (what we presume, on the basis of the best evidence, to be) laws of nature, then it seems that water changing to wine, a dead man coming back to life, etc. are not miracles because they are not incompatible with QM. (Plantinga 2008, 382)

So there are good reasons why noninterventionists look to quantum mechanics, although it might not yet be clear why such events are physically possible.

4.5.2 A Bit More of the Physics

As we've already seen (Section 2.3.3), the state of a system in classical mechanics is often represented as a point in phase space. Mathematically, these points are just ordered sets of numbers. The axes in Figure 2.4 represent the angle θ of a pendulum and the angular velocity $\dot{\theta}$ of the swinger bob multiplied by a constant. One point in the phase space represents one system state, $(\theta_1, k(\dot{\theta}_1))$.

In quantum mechanics, states can be represented in different ways, all of which are rather difficult to visualize except for very simple systems. One way is the so-called wavefunction $\psi(x)$, where x is a particular location. $\psi(x)$ is, as the name suggests, a kind of wave spread out across space. It depends on, or more precisely is a function of, the value of x. Like the waves on a lake, the "height" of $\psi(x)$ varies for different values of x. Given the wavefunction for a system, there is a rule (the Born rule) to determine the probability of that system having such and such a state when measured. This would include the probability of an electron hitting a fluorescent screen at some point across a lab or the probability that a given atom of uranium will decay within the next 5 minutes and register on a Geiger counter.

Quantum weirdness starts creeping in when we realize that all this talk of waves and wavefunctions isn't just math. In some sense, although no one seems to be able to say in precisely *what* sense, quantum mechanical systems behave like waves. Just like other waves, wavefunctions can interfere with one another, either constructively or destructively, all of which affect the probabilities just mentioned. The problem is that we never see these wavelike states. We never have any direct, observable evidence that an atom or electron becomes wavelike. What we have are the results of experiments that indicate wavelike behavior in the evolution of the system.

To understand how different this is from our everyday experience, think about shooting a BB gun at a target across the room. After each shot, we will observe a small hole in the target where the BB went through. Similarly, if we instead use an electron gun and a film target, we will observe dots left by the impact of the electron. Given what we see, one might be tempted to think of electrons as BBs, only smaller. From a vast range of experiments, we know things are not that simple. When the electrons left the gun, they were particles: they each occupied a specific place at a particular time. The same is true for when the electrons strike the target. We can see the dots on the film. Between the source and the target, however, things change. Those well-defined particles evolved in a wavelike way. They display interference effects that waves have but particles do not. The electron seems to become wavelike up until we try to isolate where it is, like having the electron strike a target. At that instant, the wavelike properties disappear and it once again looks like a particle at a specific point in space.

This is the simplest example of a ubiquitous phenomenon in quantum mechanics. The objects of everyday experience have definite properties. For example, the hands of a clock point to a particular place at each instant in time. Quantum mechanical systems usually do not have this definiteness about them. They instead have odd, wavelike properties that interact in ways we have difficulty visualizing and that we never directly observe. Nonetheless, experiments show that these wavelike states must occur between observations. All the talk about intrinsic probabilities and metaphysical randomness in quantum mechanics occurs at the transition between the weird, wavelike states and the definite, classical properties we're all familiar with. It is this instantaneous transition from wave to particle that is often called the "collapse of the wavefunction." Where will our electron strike the film? In classical mechanics, we could say precisely. Given the initial conditions and Newton's laws, a particle must behave in a particular way. A Laplacian demon could predict the outcome. In quantum mechanics, there is no precise fact of the matter where the electron will land. Given its wavefunction, there is merely a probability that the electron will land at such and such a location. We are not resorting to probabilities here because of our ignorance. The laws of physics themselves are such that *there is* nothing more than a probability that the particle will land in some location. A Laplacian demon could make no better prediction. When a wavefunction collapses, the event is to some extent ontically random.

When does this probabilistic transition from wave to particle happen? Under the orthodox (von Neumann) interpretation of quantum mechanics, the collapse of the wavefunction only occurs during "measurements"—a notoriously ambiguous term. On one standard view, measurement is something we do; people conduct experiments and take measurements. Physical devices are merely a means for doing so, but such devices themselves do not literally measure anything. People do.

This raises a number of paradoxes including Schrödinger's cat. In this famous thought experiment, a cat is left in a sealed room with a covered dish of food on one side and a bomb on the other. Next, we set up an experiment in which a single electron is fired across the room toward two detectors. There are various ways of inducing the electron into a superposition state such that there is a 50–50 chance of its wavefunction collapsing onto one or the other detector. If the collapse occurs such that Detector A is triggered, the food dish raises and the cat eats. If the collapse occurs such that Detector B is triggered, the bomb explodes and the cat dies. (As a former cat owner, I think I have a sense for why this particular animal was chosen.) Let's say that the whole apparatus was set to run automatically an hour ago. Our intuition would be that by now the cat must either be finished with dinner or in pieces scattered about the room. Recall, however, that a wavefunction does not collapse until a measurement is taken and measurements, at least on one standard way of thinking about it, are acts that *we* do. Electronic devices do not take measurements; they are merely tools people use to do so. Hence, until someone looks into the room, the entire apparatus, cat and all, are in an uncollapsed wavefunction which includes both the dead cat outcome and the live one. Both states exist until a measurement is taken and the wavefunction collapses. Before that time, there is no fact of the matter about whether the cat is dead. Not even a Laplacian demon could know.

There are other definitions of 'measurement,' but whatever it means precisely, it is clear that wavefunction collapse only occurs during these special events. Measurement is the particular junction where the odd wavelike properties disappear and the observable world of our everyday experience emerges.

There is, of course, a lot more one could say. Some philosophers of physics spend their entire careers on issues in and around quantum mechanics. For our purposes, this is far enough. When advocates of QD talk about the randomness of quantum mechanics, they are referring to the collapse of the wavefunction. There and there alone is where the metaphysical openness of quantum physics is found. Apart from the collapse of the wavefunction, quantum mechanics is completely deterministic. Without these instantaneous, random transitions, there would be no quantum indeterminism on which to base a noninterventionist theory of divine action. This is an important limitation. Quantum mechanics is not synonymous with wavefunction collapse. Many physicists do not consider the collapse of the wavefunction to be a particularly important aspect of quantum theory. Many do not even believe there is such a thing. This is where our critique begins.

4.5.3 Collapse Theories

As we have seen, measurements are unusual events in quantum mechanics. Quantum systems generally evolve according to Schrödinger's equation, which is fully deterministic. Take away measurements and the ontological

randomness in quantum mechanics we have been talking about is eliminated. This leads to serious limitations for QD.

Let's grant Murphy's claim that macro-events are in some sense composed of micro-events. She argues that God governs the macro-world by
determining the micro-realm, which is itself quantum mechanical and
indeterministic. The problem, once again, is that quantum mechanics is
only indeterministic in part. Outside of measurement events, there is no
metaphysical randomness and hence no space in which God might act. If
special divine action is limited to measurement events, then God's ability to
influence the universe at large is highly constrained: God cannot act in the
world during the normal evolution of physical systems, that is, as they
evolve according to Schrödinger's equation. Moreover, from the Big Bang
up until the appearance of sentient creatures, there were no measurements—however exactly one wants to define that term. Without measurements, the quantum mechanical evolution of the universe was fully
deterministic. Hence, God could not influence the cosmos for most of the
last 14 billion years. That would seem to be quite a limitation if one is hoping that quantum mechanics can provide a robust means of divine action.

One way to avoid some of these problems is to change the definition of
'measurement' such that a consciousness is not needed.[10] For example, one
might think that macroscopic objects, such as the detectors in the
Schrödinger's cat example, are themselves capable of measuring particles in
superposition. On this view, any time a particle in superposition interacts
with a large enough object, there is a measurement and hence a collapse
without any need of an observer. Schrödinger's cat is therefore either eating
contently or deceased well before the experimentalist peeks into the room.

A new question arises, however: What counts as a macroscopic object?
Geiger counters? Yes, those are certainly large enough. A bacterium? Well,
it's not *macro*scopic, but it is living and can absorb photons, so let's count
that as well. Now, things get tricky. We can't go all the way down to molecules, since quantum mechanics plays a key role in molecular bonding. The
objects have to be large enough to describe in classical terms rather than
quantum ones. So at what scale, between molecules and single-celled
organism, do we find macroscopic objects capable of collapsing the wavefunction? The problem is that there doesn't seem to be any *physical* reason
for drawing the line one place rather than another.[11] In fact, there is no
physical reason why the superposition of a particle can't envelop a bacterium,
Geiger counter, cat, or room. The distinction between micro- and macroscopic is driven by the limits of our sense organs, rather than anything in

the physical world. In other words, macro and micro are terms of convenience for observers like us, rather than a physical demarcation one might discover in nature. For this reason, most philosophers of physics do not believe that expanding the notion of measurement will solve the problem.[12]

One way out of these difficulties would be to find an interpretation that does not rely on measurements to collapse the wavefunction. Ideally, there would be a vast number of spontaneous collapse events in which God might act that in no way depends on people. Fortunately, there is one, the so-called Ghirardi–Rimini–Weber (GRW) interpretation (Albert 1994, 92–99). On this view, there is a small probability that the wavefunction of a particle will spontaneously collapse to one specific location at any given time. No measurement is needed. When one particle in an object undergoes this spontaneous collapse, it induces collapse on the whole. With so many particles in a macroscopic object, the collapse of at least one is virtually guaranteed every 10^{-5} seconds or less. This ensures that the objects of our experience will retain their familiar classical properties. We will never see macroscopic objects in indeterminate states, like the live cat/dead cat in the Schrödinger thought experiment. The spontaneous collapse of one particle in the cat would induce a collapse in the room as a whole. On the GRW interpretation, the cat is either dead or alive well before anyone looks in the room.

The upshot here is that there are innumerable random events on the GRW view and collapse is no longer restricted to measurement. QD is committed to such an interpretation if God is to have enough freedom to significantly influence the physical world.

4.5.4   *The Amplification Problem*

Let's assume that GRW rescues QD from the measurement problem. There are plenty of random events under the GRW interpretation in which God might act. Even so, the horizon is not yet clear.

The real issue for QD is not the number of such collapse events, but their "size." We very seldom bump up against the weirdness allowed by quantum mechanics, like the water tunneling through my glass. Physicists from Newton to Maxwell were no fools: nature *looks* classical. The unusual states allowed by quantum mechanics get washed out at the macro-level. As a rule, we only observe the wavelike effects of matter and the randomness associated with the collapse of the wavefunction in specially designed experiments.

In short, although there are many collapse events under GRW, they don't have much of an effect on the world of our experience. By analogy, conservation of momentum dictates that if one throws a pebble at the side of an elephant, it will have a physical effect, but not a noticeable one. The same is true for most quantum collapse events and the world of our experience. Such events lack the causal *oomph* to do much—*oomph* being a technical term in the philosophy of science.[13] So yes, in principle, God could use quantum events, but there isn't much that God could do with them.

For God to make use of wavefunction collapse, its effects would need to be amplified in some way, as every major proponent of QD seems to recognize:

> The question about the amplification of quantum events, for example, is crucial; if indeterministic quantum chance is entirely subsumed within higher level deterministic regularities, then it will be of no use to the theologian looking for a means of non-interventionist special divine action. (Tracy 2003)[14]

Without amplification, the effects of quantum indeterminacy do not register at the scale of our experience. Can quantum events be amplified in the manner required? There is some reason to be pessimistic:

> I am not saying that there are never circumstances in which quantum effects are amplified to have macroscopic consequences, only that they are unlikely by themselves to provide a sufficient basis for human or divine freedom. ... (Polkinghorne 1989, 28)

> According to most accounts, especially in popular literature, quantum theory is indeterministic (the "collapse of the wavepacket"). To the theologian who naturally asks whether this might afford some scope for divine action in the world I say, "Beware," ... (Butterfield 2001, 112)

The lack of amplification is the key physical hurdle for QD.

Advocates of QD do point out that there *are* macroscopic effects of quantum mechanics. Superfluidity and superconductivity are two examples (Russell 2008a, 590). Moreover, quantum mechanics is not just a matter of theoretical physics; it is essential for the existence of chemistry, for example, the Pauli exclusion principle. Even the design of computer chips sometimes requires electrical engineers to deal with quantum mechanics. Quantum effects are therefore not isolated in some inaccessible corner of reality.

Unfortunately, these examples don't help matters. No one is claiming that quantum mechanics is irrelevant to the rest of science. DNA-based life itself would be impossible in a fully classical world.[15] The issue here is the amplification of wavefunction collapse events, not whether the macroscopic realm needs a quantum mechanical foundation—it does! But as we mentioned earlier, quantum mechanics is not synonymous with the collapse of the wavefunction. Even if the wavefunction *never* collapses, as some interpretations claim, the theory of quantum mechanics would remain untouched, including phenomena such as superfluidity, superconductivity, and most of the other examples that show up in the literature. No one is questioning the truth of quantum mechanics here. The issue is whether collapse events have observable effects outside of highly engineered lab experiments.

George Ellis presents two examples that do get to the heart of the amplification question. The first has to do with photons and animal sight:

> In some species the eye can detect individual photons falling on the retina. The photon is absorbed by a molecule of rhodopsin, eventually resulting in a nervous impulse coming out of the opposite end of the cell with an energy at least a million times that contained in the original photon. The amplification of the incoming signal is due to a molecular cascade of reactions, but with much of the amplification in the initial step, where the single photon-excited rhodopsin passes on the excitation to at least 500 molecules of transducin within one millisecond. (Ellis 2001, 260)

Thus, the electrochemical nature of the mammalian eye provides an amplification mechanism for photons. A better known example deals with genetic mutation:

> A second example has been presented by Ian Percival, who states that "DNA responds to quantum events, as when mutations are produced by single photons, with consequences that may be macroscopic—leukemia for example." In this case the amplifier is the developmental process by which the information in DNA is read out in the course of the organism's developmental history. A mutation might of course have more beneficial effects than mentioned by Percival (e.g., enhanced cognition). Indeed, mutations caused by cosmic rays may well have played a significant role in evolutionary history. (Ellis 2001, 260)

This also seems to count as amplification. If cosmic rays or terrestrial radiation were to cause specific mutations, this could in principle have a long-term effect on the evolution of a species.[16]

These examples are good as far as they go, but they don't seem to go all that far. All told, God is able to use wavefunction collapse to influence photons in the eye and point mutations in DNA-based organism. That's significant, but still not the robust theory of special divine action that advocates of QD had thought it might be. QD still seems to need a mechanism of amplification, one that is not so ad hoc. Many are now pinning their hopes on chaos theory.

4.5.5   *Chaos to the Rescue?*

As we saw in Section 4.4.3, chaos theory shows how a slight change in the state of a system can produce dramatic change over time. Many QD advocates believe that chaotic systems can thereby amplify the effects of wavefunction collapse. Slight changes of state at the quantum level can bubble up into the macroscopic realm by way of the butterfly effect. Instead of a butterfly flapping its wings, the change in conditions is provided by QD. With so many chaotic systems in nature, QD-plus-chaos would seem to provide far greater scope for divine action than the alternatives.

The question now becomes this: Is there enough chaos in nature to do the job? Popular science pieces often claim that chaos is ubiquitous. This is true in some sense but is also highly ambiguous. In what way precisely is chaos prevalent?

The main answer is a matter of mathematics. Physicists often use differential equations to model natural processes. When we talk about the laws of physics, that usually means a law statement in the form of differential equations. One of the basic distinctions among differential equations is linear as opposed to nonlinear. If the sum of two solutions to such an equation is itself a solution, the equation is linear. If not, the equation is nonlinear. Most differential equations are nonlinear; the linear ones are the exceptions. Hence, if nature is governed by differential equations, as it seems, then on purely mathematical grounds, one would expect most systems to be nonlinear and chaotic.[17] "If you draw a curve 'at random' you won't get a straight line. Similarly, if you reach into the lucky dip of differential equations, the odds against your emerging with a linear one are infinite" (Stewart 1989, 83). Likewise, there are many more irrational numbers than rational ones. Intuitively, if one were to pick a real number at random, the odds would be (literally) infinitesimal that it would be rational. The same is true for drawing a linear equation at random from the space of all such equations. Physicist Roland Omnès puts it this way: "From the standpoint of a mathematician, there are many more chaotic systems than regular ones. This means that, if one were to generate the Hamilton function at random, the chances would be very high that one would get a chaotic system" (1994, 230). Since there is no reason for nature to prefer linear over nonlinear systems, with all likelihood, most real-world systems are non-linear and chaotic. This is the Argument for Ubiquitous Chaos.

Is this argument sound? On purely mathematical grounds, yes. In the space of differential equations, almost all of them are nonlinear. However, as Omnès goes on to say, "[N]ature does not play that kind of game and the majority of ordinary objects around us are not chaotic, except maybe at a very small scale." The point is that, as all theoretical physicists know, arm-chair mathematical reasoning does not necessarily carry over into real-world systems. The prevalence of chaos in nature is an empirical matter and cannot be determined from measure-theoretic arguments about nonlinear equations.

The Argument for Ubiquitous Chaos is also too strong in at least one way. If the analogy to the real numbers is as close as it seems, then physicists should never find linear equations governing natural systems. Unless nature is somehow biased toward linearity, then the odds of finding a realistic, linear law are infinitesimal. To find one useful linear equation would be so unlikely that it would cry out for explanation, much like fine-tuning. But, of course, physics has discovered such laws. Consider this: Schrödinger's equation—which, if anything, counts as a fundamental law of nature—is linear.

A completely different argument is given by one of fathers of modern chaos theory, physicist David Ruelle. The title of Ruelle's *Physics Today* piece, "Where Can One Hope to Profitably Apply the Ideas of Chaos?" (1994), was puzzling given the rate at which books and articles on chaos had been produced in the previous 15 years. One might think that the answer is obvious: "everywhere." Among other things, Ruelle argues that chaos is not everywhere, at least not the way the claim is usually interpreted. Chaotic dynamics is much like noise: in a given system, there might be a little or a lot. If the chaotic component is small relative to the overall behavior of the system, its presence will have little or no effect.

Consider a simple analogy. I once saw one of my sons ride his tricycle around in a small circle. On each pass around the circle, his wheels never went over the exact same path as before. The claim 'Christopher is riding in a circle' is accurate but not precise. If one examines his trail up close, there is a good deal of variation. On a large scale, the motion is regular and periodic; on a finer scale, each lap is unique. The point is that although the path of the tricycle is irregular, this does not entail that the path is completely haphazard. The imperfect, random component on each pass is slight compared to the overall circular path.

Likewise, to say that a given system is evolving chaotically often means merely that there is a small, random-looking component in the background of a completely regular time series.

For another illustration, let's say that during a telephone conversation I detect a slight hiss in the background. The hiss might be due to thermal effects in the telephone lines, but it might also be due to deterministic chaos in the network. A dynamical systems analyst may be interested in discovering which is the case, but as for me and my call, it doesn't matter. The hiss is barely detectable. A small amount of background chaos—or background noise—only has a slight influence on the audio quality. The voice harmonics of our conversation dominate the signal. Crudely put, the dynamics is chaotic, but not much.

Real-world chaos is often limited in a similar way. Let's grant for the moment that chaos is everywhere in nature. In many cases, it is present only on the fringes and has little effect on the behavior of a system. Sometimes this is obviously the case. When researchers at the Harvard Medical School argue that heartbeats are chaotic (Ruelle 1994, 26–27), they clearly don't mean that healthy heartbeats are completely erratic. The point of the tricycle, telephone, and heartbeat examples is that the mere presence of chaos in a dynamical system does not entail wholesale disorder. Chaos often shows up only in the background of an otherwise regular evolution. It comes in degrees, just like thermal noise.

Now, recall why QD advocates want to add chaos to the mix: a small change at the quantum level is supposed to produce a large change in the evolution of a chaotic system. This is true, in principle. In reality, the amount of amplification provided by chaos is small. Consider the telephone example again. Say that God exercises complete control over the chaotic portion of the signal via QD. Unfortunately, this would have no noticeable effect on the conversation, and so the extent of God's influence over this system is quite limited. If most real-world chaos is likewise restricted to the periphery, then QD-plus-chaos cannot provide significantly more leverage for God's action than QD alone does.

The bottom line is that the mere presence of chaos in a physical system might not amount to much. 'Chaos' is a highly suggestive term. It tends to imply turmoil, disorder, and unpredictability. The truth is somewhat disappointing. Chaos comes in degrees and is often found in the midst of stable structures (e.g., convection cells) and dynamics that are predominantly regular (e.g., heartbeats). The fact that there is some chaos in the atmosphere does *not* mean a butterfly outside my window can change the weather in China. The same goes for quantum fluctuations in a chaotic system.

QD needs an amplifier and chaos was its best bet. Unfortunately, chaos is far more limited in nature than popular discussions make it out to be. It therefore cannot serve as an effective amplifier of wavefunction collapse events. QD's amplification problem remains unsolved.

While this is not a fatal flaw, there is a more fundamental problem lurking. Should a theist really want this in the first place? In some circles, non-interventionism is beyond question. Rival theories of divine action are simply dismissed, often with contempt. Let's now consider whether such an attitude is warranted.

4.6   Noninterventionism: Goring the Sacred Cow

What I am about to argue will keep me from being invited to certain conferences in the future. Challenging a view that is a settled question in some circles invites a harsh response. Nonetheless, I want to go back to the beginning of this chapter and reconsider the arguments for noninterventionism one by one.

4.6.1   *The Infinite Clockmaker*

The image of God as an infinite clockmaker started in the heyday of mechanical philosophy. If God is a sort of perfect craftsman, then it seems incongruent that God's creation would need maintenance, like my very imperfect lawn mower. An infinite clockmaker would make a clock that did not need to be adjusted every so often. That is the image that motivates noninterventionism, but what exactly is the *argument* here? Why is this sixteenth-century picture the dominant one?

Consider another. I have a friend, David, who is restoring a Milwaukee Automobile Co. Model A steam car (Figure 4.1). Why, one might wonder, would someone want to do such a thing?

After all, modern cars are better, safer, and more reliable in every way imaginable. Spending time and money to recreate a technologically obsolete vehicle seems like a waste of time. The answer, of course, is

Style A

Standard, - - - - $750.00
F.O.B. Milwaukee, Wis.

Capacity, - - Two passengers
Wheels, - - - - - 28-inch
Tire, - - - 3-inch pneumatic
Gasoline Tank, - - 5 gallons
Water Tank, - - 21 gallons
Weight, empty, - - 700 lbs.
Weight, full tanks, - 900 lbs.
Seat, - - - - Spindle back
Tread, - - - - 4 feet 6 in.

Style B

has 6-gallon gasoline tank and
30-gallon water tank, with wider
body, and 41-inch seat in the
clear; stick seat instead of
spindle.   Price, $850.00.

EQUIPMENT

Rubber Storm Apron, Rubber
Mat, Side Lamps and Covers,
Odometer, Torch and all neces-
sary tools.

Figure 4.1   Milwaukee Automobile Co. Model A.

that David finds joy in the ongoing restoration. He looks forward to the day when the car is finished and he can drive it around and share the experience with others. People who restore old cars often have the money to pay someone else to do the work or they could buy the completed car outright, but they choose not to. The restoration process is part of what they enjoy. A similar idea applies (so I'm told) for those who like gardening.

Why think that the theistic God would not directly interact with creation, as if it were sitting on a divine mantle never to be touched? Perhaps noninterventionists are using the wrong metaphor. To be fair, advocates of QD do allow for divine action in nature so long as it falls within indeterministic gaps. But why are *those* boundaries to divine action inviolable? Say that to make the Model A legal to drive on public roads, David has to alter the original design somewhat. He must add license plates and a mirror, violating his intention to restore the car according to the original specifications. Some sort of trade-off must be made. He must either alter the design slightly or keep the car off the road. It is likely that God has prioritized goals at times, much like a parent does for a child. I never want my children to be in pain, but I do want to have them inoculated from disease. It is not a sign of imperfection that God must make choices. One of those choices might be to suspend a natural law.

4.6.2   *Capricious or Inconsistent*

The Greek gods were not so different from us. They displayed a full range of virtues and vices, often with no reason for choosing one way rather than another. The theistic God is not supposed to be like that. However God acts, it is not capricious—performing miracles on behalf of some and withholding them from others for no reason.

Assume for the moment that there really are miracles and that these violate the laws of nature. Say that my student's brain tumor was directly healed by divine intervention. Now, say I have another student this year who, as near as I can tell, is in exactly the same circumstances as the first: age, socioeconomic background, religious piety, etc.—identical. The second student goes to the same group for prayer and yet this student is not healed. God willed the healing of the first but not the second even though their situations are the same. Was this a capricious act?

Well, from our point of view perhaps. But the fact that we cannot discern a difference does not mean there isn't one. Very few people can tell the difference between earrings made of cubic zirconium and those made of diamond, but there is a difference nonetheless. Just because we fail to see a pattern or reason for why God might intervene in one case but not the other does not entail that God has no reason. If we were able to discern matters from a God's-eye perspective, the healing of the first student might strike us as completely reasonable and not at all capricious. This is not a new idea. Medieval voluntarists recognized a distinction between ordinary and extraordinary providence but "considered this difference to be a function of incomplete human knowledge" (Harrison 2002, 79). They held that miracles take place according to a preordained and lawful pattern, albeit a pattern that is difficult to discern. To put it another way, the inability to fit miracles into a pattern of natural law was a consequence of limited human understanding, not a difference in the mode of divine operation. The lesson here is that one should be careful about attributing capriciousness vis-à-vis divine action given our limited knowledge.

A closely related concern is that God would be inconsistent to first declare the laws of nature and continually uphold them and then to choose to violate those same laws. The first choice, some noninterventionists argue, would be in conflict with the latter.

In my view, this worry is rooted in a rather simplistic view of ends and the will. I have a general desire to not be in pain, but I have participated in sports where there is a risk of injury and even played while injured. Was I being inconsistent? Was there conflict of will here regarding pain? No, it's just a matter of circumstances. One's general will is *ceteris paribus*—all things being equal—but there are exceptions. My general will is to keep my kids away from explosives, and yet I allow them to light fireworks at certain times. Conflict of will? Inconsistent? I don't see why. The notion of will is complex and layered. God can have a general will to govern the universe via the laws of nature and yet allow for special exceptions. Nothing here constitutes a "conflict of will."

4.6.3   The Problem of Evil

The existence of evil in the world is a perennial problem for theism. The question in terms of divine action is, if God steps in and prevents some evil, why not more? Why not all? A theology in which God seldom (or never) intervenes allows evil to be chalked up to free will and the working out of the laws of nature. The murderer (mis)uses his freedom and is therefore responsible, not God. As for natural evil, any world with stable laws will produce situations in which people will be hurt. The rain cycle is good, for example, but swimmers will sometimes drown. That type of evil is merely a matter of people getting in nature's way.

Noninterventionists argue that if God sometimes prevents natural evil by suspending the laws of nature, there is no principled reason why God does not prevent more.[18] They believe that the problem of evil is therefore worse under a traditional view of divine action. The only way to account for natural evil is to give nature a kind of autonomy analogous to human freedom. Just as God does not intervene to prevent moral evil, God does not intervene in nature, all for the sake of autonomy in creation.

The problem of evil is the weightiest objection to theism. It would be quite a payoff for noninterventionism if it could account for natural evil. Unfortunately, it does not. There are two reasons. First, it seems that God could have made different laws that would produce less evil in the world. Natural evil might be inevitable in a law-governed universe, but some natural hazards would have been eliminated had the laws been different. Hence, even if we can account for the existence of evil, the amount of evil in the world is still an issue for every model of divine action. Second, if under QD God is supposed to have influenced evolution by using radiation to tweak genomes, why didn't God make us more resistant to cancer? Why didn't God make mosquitos with a dislike for human blood? And so on. You get the idea.

This is enough to show, I believe, that there is no easy way out of the problem of evil. Neither QD nor any other form of noninterventionism escapes.

4.6.4   *Intervention Conflicts with Science*

Many claim that science is committed to methodological naturalism: explanations must be restricted to naturalistic causes. (We will consider this in more carefully in the next chapter, but let's just grant it for now.) Miracles are therefore contrary to good science. Insofar as our thinking is scientific, then, miraculous interventions must be dismissed as the archaic thought of prescientific peoples.

This is a common inference in the divine action literature. Note, however, that methodological naturalism merely says that science cannot appeal to God as an explanation for any event. That's not the same as science forbidding divine action, miraculous or otherwise. All that methodological naturalism would entail is that *if* there are miracles, science might bump up against events for which it cannot offer a true explanation. Strictly speaking, special divine action is at worst nonscientific, not unscientific. Methodological naturalism insures that special *divine* action will never show up as part of a scientific explanation.

Where, then, is the conflict with science? Not even a creationist—an arch interventionist if there ever was one—necessarily wants miracles to be included in a scientific theory.[19] This shows that one can hold to a strict methodological naturalism and still believe in miracles. My student with the brain tumor did not expect his doctors to put "miraculous healing" in his medical records. Presumably, the physicians treated the event as a medical mystery; there is no medical explanation. If methodological naturalism is merely trying to keep miracles out of scientific discourse, this can be done without a blanket rejection of the miraculous. There are truths beyond the reach of science, but there is nothing unscientific about that. History contains truths that are not a part of science. That doesn't make history unscientific.

That miracles are beyond the domain of science is not a problem. The only worry for theists would be if somehow science shows that supernatural interventions cannot happen. For there to be a conflict, interventions must be unscientific in the sense that perpetual motion machines are unscientific: forbidden by physics itself. But since methodological naturalism is merely a shaping principles and not part of any specific theory, it isn't clear how science could even possibly show that God cannot intervene in nature (more on that line of thought in Section 4.7).

4.6.5   *God of the Gaps*

The objection is that God ought not be used to fill explanatory gaps in our knowledge. When this has been done in the past, science eventually closes the gaps and God gets squeezed out. Hence, using God as a theoretical posit is a losing strategy for theism and should be avoided.

There is general agreement on this point. Few think that we should posit God as an explanation for every piece of surprising or unexplained data. Whether God should *ever* be used to explain a natural phenomenon is just the question of interventionism in different garb. If God does intervene in nature, then the only true explanation for some events would require God as a part. If God never intervenes, then the true explanations would never mention God. Which is it?

To resolve this, we must consider more closely the notion of divine intervention in a physical system.

4.7   Intervention and Determinism

There is a great deal of fear and loathing over determinism in the divine action literature. Consider this quote by Saunders again:

> [In] a totally deterministic world … the causal nexus of science is drawn so tight that there is no real freedom for either God or human beings. In such a world Laplace's famous demon reigns supreme. God cannot act in any creative way through the causality of science and still remain true to the deterministic rules put in place at creation. (2000, 254)

Noninterventionism often hinges on finding indeterministic gaps in which God might act. Quantum mechanics seems to be the best candidate since, as everyone knows, classical mechanics was deterministic.

But what if that bit of conventional wisdom isn't true? What does it mean to say that classical mechanics was fully deterministic?

'Determinism' is an ambiguous word. The view presupposed throughout this discussion so far can be traced back to the Dutch philosopher Baruch Spinoza (1632–1677): "Nothing happens in nature which does not follow from her laws" (*Theologico-Political Treatise* Bk 2 VI:21). In that kind of world, there is no contingency whatsoever and no space in which God might act. The discovery of quantum mechanics broke the grip of determinism for the first time, or so the story goes.

In fact, Spinoza's view does not even hold for classical mechanics. Consider Mark Wilson's train wheel example (1989, 509). As the piston moves back and forth, the wheel rotates counterclockwise (Figure 4.2).

The motion of the piston determines the motion of wheel—precisely the sort of determinism that Spinoza had in mind. Now, consider a case when the train just happens to come to rest in the configuration shown in Figure 4.3. As the piston moves to the right, will the wheel move clockwise or counterclockwise? If the piston supplies enough force, the wheel must rotate. Which way will it go according to the laws of classical mechanics?

Surprisingly perhaps, there is no fact of the matter. The wheel must turn (assuming nothing breaks), but how it will rotate is not determined by the laws of nature. More than one effect is possible from a single cause. The outcome is indeterministic.

Consider another of Wilson's examples, the Euler strut (Figure 4.4). A perfectly homogeneous, symmetric strip of metal is pinned at both ends

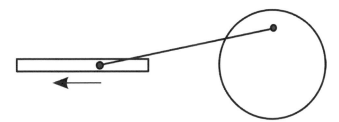

Figure 4.2   Train wheel in motion.

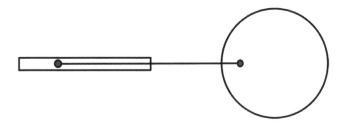

Figure 4.3   Train wheel at critical point.

Figure 4.4   Euler strut.

and pointing straight up. In addition, let's put the whole apparatus in a vacuum and isolate it from outside vibrations. We then start putting a balanced load directly on top. If we keep increasing the load, the strut must eventually buckle. A thin strip of metal can only support so much weight. Will it buckle to the left or right? Again, there is no fact of the matter according to classical mechanics. The laws and the initial conditions are not sufficient to determine the outcome. Laplace's demon cannot know in what manner the strut will collapse.

There are other examples including domes (Norton 2008), indeterministic fractures, collisions, and turbulence, many of which are explained by Earman (1986). The strut and wheel are enough to show that if we understand determinism to mean that every event is fixed by the laws of nature, then not even classical mechanics—that supposed paradigm of physical determinism—counts as fully deterministic.

So it seems that quantum mechanics did not break the absolute grip of determinism after all, as noninterventionists often suppose. Determinism in physics is more nuanced than conventional wisdom might suggest.[20] Spinoza's view, reflected by Saunders, is oversimplified. Newton himself would not have endorsed it. In fact, determinism isn't the issue at all, or at least it should not be. Noninterventionists have misidentified the enemy.

Let's ignore the Euler strut and other complications for the moment. Say that our universe is Newtonian: every system is mechanistic and obeys Newton's laws. The grandfather clock in our living room, like everything else, works via purely mechanical interactions. There are no gaps in the clock's dynamics. A Laplacian demon would not break a sweat. With all that, nothing in this picture entails that a child cannot stick his finger in and spin the hands backward. The dynamics of the clock have to respond to a new impressed force. Now, does the finger-in-the-clock addition violate any natural laws, for example, the conservation of energy? After all, an engineer working out the relevant equations for the clock would clearly assume that mass–energy is conserved for that system. The answer, of course, is no. Conservation principles do not forbid outside action; they say what happens to systems that are considered closed or isolated. If something mechanically interferes with a "closed system," all appeals to conservation are off.

The fact that the clock is deterministic in no way rules out external forces. We can treat the clock as an isolated system for the sake of analysis, but in reality, the very notion of an "isolated system" is an idealization. In a world with gravity and electromagnetism (infinite range forces), there are no isolated systems. Every bit of matter influences every other bit.[21]

This all makes perfectly good sense when it comes to clocks and kids. No one at the height of the old mechanical philosophy would have had any conceptual difficulties with this example. Let's now take the child out of the picture and put in a supernatural cause. Say that God creates an impressed force on the hand of the clock applying the same torque as the child's finger. What is the difference between the two examples? One is the nature of the agent involved, natural versus supernatural, but I don't see how physics would have anything to say about that. Neither the clock nor the laws that govern its operation care about the source of an impressed force. The same free-body diagram (which captures the forces and mechanical interactions) will apply to any force applied to the hand of the clock: contact forces or electromagnetic, forces generated naturally or supernaturally. God has merely changed the conditions under which the deterministic laws governing the clock operate. God has not violated or suspended those laws any more than the child.[22]

What this shows is that there is nothing about determinism that would prevent God from acting in the world. In the debate over divine action, determinism is a red herring.

Let's briefly consider a few objections to this picture. First, one might grant that the child's interference invalidates the conservation of energy with respect to the clock. Conservation principles only hold when a system is isolated from outside forces. But surely God's influence on the universe as a whole must violate energy conservation. When God moves the hands of the clock, energy is being created within our spacetime, thus a violation.

The reply to this objection is straightforward: the conservation of energy does not apply to the universe as a whole. This is a consequence of general relativity (Wald 2010, 70). In an expanding universe like ours, there is no widely accepted definition for the energy of the universe as a whole and hence energy conservation cannot be violated (Carroll 2010, 87–88). More precisely, energy conservation is tied to a symmetry in the laws of nature known as *time-translation invariance*. Since this symmetry does not hold in GTR, neither does the conservation of energy. A law cannot be broken if it does not apply.

Second, one might object that the child is also a physical system, like the clock, but God is not. Hence, there is an important disanalogy between the two. One cannot merely replace the child's finger with God's will in this example.

There are certainly important differences between the two cases, but not when it comes to determinism. We stipulated that in a Newtonian world, the clock is a deterministic system. The child does not violate any laws of nature by moving its hands. God is a different kind of agent, no doubt, but again the *source* of the newly introduced force is irrelevant to the physics. So long as the hand of the clock experiences a torque, the equations work just the same regardless of how the force originated. There might be other metaphysical questions about how an immaterial being can affect a material system, but those have nothing to do with the issue of determinism. We might not have a clear and precise understanding of how God can create the impressed force on the clock hand, but we don't have any clearer picture of how God collapses the wavefunction in QD. If there is an element of mystery in my clock-intervention example, the same mystery is found within QD.

Third, what about the fact that God is, in some sense, a nonlocal source? God has to step into the physical world in a way the child does not—he's already here. Within a Newtonian universe, a Laplacian demon could predict the behavior of the particles that compose the child's body along with every other particle in the universe. God's activity is different. The Laplacian demon could not predict God's intervention.[23] Doesn't that break the analogy?

Actually, no, but the question points to an important issue lurking here: determinism by itself has nothing to say about whether a system is closed to outside influences. As Earman shows, classical mechanics allows particles to enter our local space from infinitely far away. In a Newtonian world, it is physically possible both for particles here to zoom off to spatial infinity in finite time and for particles to appear from spatial infinity. "Thus, in

Newtonian space-time … [the universe] is not automatically 'closed' in the operative sense to outside influences" (Earman 1986, 34). A Laplacian demon would not be able to predict the arrival of such particles, yet no laws of nature would be broken by such an event. The laws would simply make room for the particles once they were here. The point is this: locality, closedness, and determinism are distinct matters in physics.

In sum, deterministic laws of nature in no way forbid God's direct causal influence. The clock example does not involve a violation of nature law. As philosopher and atheist J.L. Mackie put it, "The laws of nature, we must say, describe the ways in which the world … works when left to itself, when not interfered with" (Mackie 1982, 19–20). That seems right, whether we're talking about divine agents or not. If I lift a ball off the ground, I have not violated the law of gravity. Likewise, God can change the conditions under which the laws—including deterministic laws—will act without violating them. That isn't the way Spinoza understood 'determinism,' but his view is

> contrary to the modern conception of determinism according to which laws allow for contingency in "initial conditions" and necessitate only conditionals of the form "If the initial conditions are such-and-such, then the state at a later time will be so-and-so." (Earman 2000, 9)

Too many noninterventionists hold a view of determinism that physics jettisoned over two centuries ago.

In order to rule out divine acts like the clock example, one would need a principle that somehow prevented outside influences from reaching into our observable universe. This leads to the real issue. Determinism is irrelevant. The question is whether the universe is a closed or isolated system. Noninterventionists believe that science entails the *causal closure of the physical*. Roughly, the idea is that physical effects only have physical causes. Physical events can only be caused by earlier physical events in conjunction with the laws of nature.[24] If there are other types of causes somewhere in the matrix of reality, closure says that they do not bring about physical effects. God moving the hands of a clock violates the laws of nature insofar as it violates causal closure.

Having finally isolated the crux of the matter, it raises some interesting questions. For one, has any scientific observation or theory established that the universe is causally closed?

No, and it isn't clear how this could be done.[25] Closure is another metatheoretic shaping principle. Like the others, this principle is a philosophical view about the nature of reality that science relies upon. Closure is not entailed by any scientific theory, including those in physics, as philosopher Robert Bishop has argued:

> Physics does not imply its own closure. … Rather, CoP [the causal closure of physics] is a metaphysical doctrine. Indeed, physics tells us what happens when particular forces are taken into account, but nothing about what happens when influences unaccounted for by physics are present. (Bishop 2011, 606)

If no scientific theory entails causal closure, where does it come from?

Closure is imposed on many philosophical debates as a doctrine dictated to us by science: this is the way things are in nature. Notice, however, that closure in this sense is a very broad metaphysical principle that far outruns methodological naturalism. If causal closure is a metaphysical thesis, it is only slightly weaker than *metaphysical* naturalism. In other words, once closure rules out the existence of nonphysical causes in the universe, it collapses into metaphysical naturalism for causation. Closure allows for a supernatural reality but only as long as it is causally isolated from this one. It says that nonphysical causes—whatever they might be—do not produce physical effects. Only physical causes do that.

All that's standard fare for philosophers trying to play out the implications of naturalism, but why should theists agree to causal closure? Why should anyone who is not already committed to metaphysical naturalism? Noninterventionists seek views of divine action that are consonant with modern science. Very well, but as we've seen, modern science—for that matter, science since the discovery of calculus—does not entail causal closure, as Plantinga rightly argues:

> There is an interesting irony, here, in the fact that the hands-off [=noninterventionist] theologians, in their determination to give modern science its due, urge an understanding of classical science that goes well beyond what classical science actually propounds. Hands-off theologians can't properly point to science—not even to eighteenth and nineteenth-century classical science—as a reason for their opposition to divine intervention. What actually guides their thought is not classical science as such, but classical science plus a metaphysical add-on—an add-on that has no scientific credentials and goes contrary to classical Christianity. (Plantinga 2008, 380)

As is often the case, what some put forward as science is actually philosophy in a trench coat and sunglasses.

A better approach would be to take causal closure as a methodological principle for the development of science.[26] Closure shows how science should normally be pursued, what Bishop calls a "typicality condition": "[Causal closure] itself does not rule out nonphysical interventions—it only says what happens in their absence (or when they make negligible contributions)" (Bishop 2011, 607–608). This seems exactly right. In the absence of causes outside of physics, physical events will proceed via the laws of nature. This means that science, even deterministic science under methodological closure, in no way constrains special divine action.[27] One won't expect to find miraculous events among our best scientific theories, and that's fine. Truth is not wholly contained within science. Theology and physics are different disciplines.

In the end, the search for indeterministic causal gaps through which God might act was poorly motivated in the first place. So long as the universe constitutes an open system, the laws of nature do not constrain God's actions, regardless of whether determinism holds. Science and metaphysics leave plenty of room for the creation and sustaining of the cosmos and special divine action as well.

We should note in closing how conventional wisdom about the laws of nature has changed since the late medieval and early modern period. As we saw in Chapter 1 (Sections 1.2.2 and 1.2.3), the early moderns began rejecting Aristotelian essences because they believed that God had no need of intermediaries to govern nature. Descartes, Boyle, and Newton were the norm, believing that God directly and immediately rules creation through its laws. God decrees how natural entities will behave and they do so. Scientists in this period took their law statements to be the discerned patterns of God's regular actions. Miracles were nothing more than surprising ways in which God acts, but the mechanism was all the same.[28]

Today, in contrast, noninterventionists seem to view the laws of nature the way Boyle and Newton thought of the essences they had come to reject: as intermediaries that God has put into place to govern the universe (see Section 1.2.2). Noninterventionists have adopted the naturalistic view that laws are foundational for the autonomy of nature. They take it that God created and sustains the universe, but the laws do all the heavy lifting. This is an important—and in my view, unfortunate—philosophical shift. It is a philosophical rather than a scientific change insofar as modern science does not entail that the universe is a closed system or that the laws actually *govern* the universe (Section 4.3).

I have not presented a new model of special divine action here. My goal was to lift the artificial constraints placed on divine action over the last half century. Of the views presented in Section 4.4, I take Polkinghorne's to be closest to the mark, which is not to say that Sir John would endorse my arguments. He would almost certainly be concerned that my finger-in-the-clock example is too episodic to represent ongoing divine action. Perhaps so. On the other hand, if the voluntarists were right and what we think of as miracles are brought about by the same means God uses to otherwise govern the universe, there might be ways of ironing out the differences. The universe *is* intrinsically open, perhaps in ways that we have not discovered. Polkinghorne believes that physics will eventually uncover a new mechanism in which this openness is expressed. In my view, no new mechanism is needed once we understand that determinism does not entail that nature is a closed system. God would be able to interact with the universe without breaking the laws of nature even if classical mechanics were true.

In the end, the traditional view of divine action probably *is* wrong; God does not violate natural law. The mistake is not, however, with the belief that miracles can or do occur. The error is in tacitly thinking that the world is a closed network of interlocking causes that God must break in order to act.[29] As we've seen, the idea that modern science entails such a view is simply false. No scientific theory demands that nature be a causally closed system. The universe might be deterministic, as even some interpretations of quantum mechanics allow, but that would fall short of causal closure. If the universe is an open system, then agents not in that system can influence its evolution. Metaphysical naturalists and theistic noninterventionists are free to argue for causal closure, and *some* sort of argument would certainly be nice.[30] As it is, closure is generally imposed on the discussion under the assumed imprimatur of science. But science is more than natural law and natural law requires a supporting cast of initial conditions (and more). Neither science nor its laws rule out divine action, even the kind of action that makes noninterventionists uncomfortable.

Spinoza would not be pleased with this conclusion. I take comfort in the fact that Newton is on my side.

Notes

1   Philosophers recognize various kinds of possibility. To be precise, what is at issue here is *physical possibility* or *nomological possibility*: those events consistent with the laws of nature in our universe.
2   Robert Russell's view is a bit more nuanced than most insofar as he accepts an interventionism view of miracles but believes that particular mode of divine

action is atypical. He has three categories: providential nonintervention, miraculous intervention, and special divine action without intervention (2008b, 128–129). The third category will be spelled out more fully in Section 4.4.

3 The free will defense is essentially that since we have libertarian freedom, humans are responsible for the choices they make. When a murderer decides to kill, God is not culpable for the evil act, the murder is.

4 "For while Comets move in very excentrick Orbs in all manner of Positions, blind Fate could never make all the Planets move one and the same way in Orbs concentrick, some inconsiderable Irregularities excepted, which may have risen from the mutual Actions of Comets and Planets upon one another, and which will be apt to increase, till this System wants a Reformation" (*Opticks*, Query 31).

5 "Whence is it that Nature doth nothing in vain; and whence arises all that Order and Beauty which we see in the World? To what end are Comets, and whence is it that Planets move all one and the same way in Orbs concentrick, while Comets move all manner of ways in Orbs very excentrick; and what hinders the fix'd Stars from falling upon one another? How came the Bodies of Animals to be contrived with so much Art, and for what ends were their several Parts? Was the Eye contrived without Skill in Opticks, and the Ear without Knowledge of Sounds? How do the Motions of the Body follow from the Will, and whence is the Instinct in Animals? Is not the Sensory of Animals that place to which the sensitive Substance is present, and into which the sensible Species of Things are carried through the Nerves and Brain, that there they may be perceived by their immediate presence to that Substance? And these things being rightly dispatch'd, does it not appear from Phænomena that there is a Being incorporeal, living, intelligent, omnipresent, who in infinite Space, as it were in his Sensory, sees the things themselves intimately, and throughly perceives them, and comprehends them wholly by their immediate presence to himself: Of which things the Images only carried through the Organs of Sense into our little Sensoriums, are there seen and beheld by that which in us perceives and thinks. And though every true Step made in this Philosophy brings us not immediately to the Knowledge of the first Cause, yet it brings us nearer to it, and on that account is to be highly valued" (*Opticks*, Query 28).

6 Henri Poincaré is credited with solving the stability of the solar system question in the late 1800s, although Karl Weierstrass found an important flaw in Poincaré's solution before it was published. See Diacu and Holmes (1996) for an interesting history of the stability question in celestial mechanics.

7 Perhaps they should, as Del Ratzsch has argued (1987).

8 This is Saunders's point in Saunders (2002, 62), although it doesn't seem to be well understood (Tracy 2008).

9 The birth of chaos theory is usually traced to the work of mathematician–meteorologist Edward Lorenz in the 1960s, although the underlying mathematics was pioneered much earlier by Henri Poincaré and Pierre Duhem.

10 Russell holds a view similar to this.

11 See Bell (2004, 124) for a clear explanation for why this is the case.

12 David Albert is quite harsh when it comes to this and similar proposals: "There is ... an astonishingly long and bombastical tradition in theoretical physics of formulating these sorts of guesses about precisely when the collapse occurs in language which is so imprecise as to be ... absolutely useless. Some of the words that come up in these guesses (besides *measurement* and *consciousness* and *macroscopic*) are *irreversible, recording, information, meaning, subject, object,* and so on" (1994, 84).

13 Note to students: that was a joke.

14 Also see Tracy (1995, 317–318). Murphy and Russell discuss the problem in Murphy (1995, 357) and Russell (2008b, 158).

15 This is true for many reasons. One is that atoms with a positively charged nucleus and negatively charged electrons would not be stable in a fully classical world.

16 Russell takes this to be highly significant insofar as it refutes the idea that evolution is in conflict with theism is intrinsically atheistic (2008b, chap. 6). Others argue that directed evolution is not compatible with neo-Darwinism whether there is a violation of natural law or not. Darwinian mutations are random precisely in that "they do not occur according to the needs of their possessors" (Ruse 2012, 623). If God were to cause mutations to ensure that humans evolve, it would be nonrandom and hence non-Darwinian. As Ruse points out, when Darwin's friend and supporter Asa Gray first proposed a version of theistic evolution, Darwin argued that it was incompatible with his theory.

17 Strictly speaking, that a system is correctly described by nonlinear differential equations is a necessary but not sufficient condition for chaos, but those details don't affect my argument here.

18 Russell is an exception. He recognizes that any sort of divine action, interventionist or not, will tend to exacerbate the problem of evil (private correspondence).

19 *Creation science* seeks to take creationism beyond the theological realm and push it into science proper. My point is that one could be a young earth creationist for theological reasons and yet not think that such a view is properly part of science itself. Such a creationist would interpret geology, cosmology, and other historical sciences in an antirealist manner.

20 "Classical determinism is not the mummified relic that philosophical literature portrays it to be, but a living and breathing creature capable of generating surprising twists and turns" (Earman 1986, 53).

21 A closed system is not necessarily one that is isolated from its environment. The air molecules in a piston constitute a closed system, but one that is capable of receiving heat energy and doing work on the piston head. Classically, closed systems can exchange heat but not matter. See Larmer (forthcoming) for a related view of divine action, closed systems, and the conservation of energy.

22 Plantinga makes the point differently, but it fits nicely with what has been said here (2008, 374–375). Also see Alston (1993, 190).

23 This objection was suggestion by Thomas Tracy (private correspondence).

24 We should note that Russell allows for exceptions to this rule. He believes that there are miraculous events in which God directly intervenes, but these are always observed and religiously significant (private conversation). For example, Jesus raising Lazarus from the dead would be a miracle. These are, by definition, unusual events. God's normal means of action is within the unobservable and indeterminate spaces allowed by nature, most especially quantum mechanics.

25 Russell thinks otherwise: "As we know, in classical physics, nature is a closed causal system describe by deterministic equations" (2008b, 157). As I argue here, we know no such thing. Closure is a metaphysical add-on to classical physics. Neither absolute determinism nor closure is entailed by classical mechanics or electrodynamics.

26 Thomas Tracy made a proposal along these lines (2008, 601).

27 Philosopher–theologian F.R. Tennant made the same point nearly a century ago (1924, 384).

28 For more on the notion of law and divine action from Newton to Darwin, see Brooke (1992).

29 After writing this chapter, I discovered a highly sympathetic view held by the philosopher of biology Elliot Sober. "What I want to consider … is the view that God *supplements* what happens in the evolutionary process without violating any laws. An intervention, as I'll understand the term, is a cause; it can trigger an event or sustain a process. Physicians do both when they intervene in the lives of their patients. Physician intervention does not entail any breakage in the laws of nature; neither does God's" (2011, 362). Sober understands those who believe that God guides the evolutionary process to be taking a "hidden variables" approach to the theory, something akin to what Einstein thought of quantum mechanics.

30 The best I have seen is Papineau (2009). In Section 2.4, Papineau looks at the history of physics relating to conservation principles and causal closure. In the end, however, he merely appeals to what most scientists believe today. As I argued earlier, there is a considerable gap between scientists using closure as a metatheoretic shaping principle and the claim that science entails the truth of causal closure.

References

Albert, David. 1994. *Quantum Mechanics and Experience*. Cambridge: Harvard University Press.

Alston, William P. 1993. "Divine Action, Human Freedom, and the Laws of Nature." In *Quantum Cosmology and the Laws of Nature: Scientific Perspectives on*

Divine Action, edited by Robert Russell, Nancey Murphy, and C.J. Isham, 2nd ed., 185–206. Berkeley: Center for Theology and the Natural Sciences.

Bell, John Stewart. 2004. *Speakable and Unspeakable in Quantum Mechanics: Collected Papers on Quantum Philosophy*. Rev. ed. Cambridge: Cambridge University Press.

Bishop, Robert. 2011. "Free Will and the Causal Closure of Physics." In *Visions of Discovery: New Light on Physics, Cosmology, and Consciousness*, edited by Raymond Chiao, 601–611. New York: Cambridge University Press.

Brooke, John Hedley. 1992. "Natural Law in the Natural Sciences: The Origins of Modern Atheism?" *Science & Christian Belief* 4 (2): 83–103.

Butterfield, Jeremy. 2001. "Some Worlds of Quantum Theory." In *Quantum Mechanics: Scientific Perspectives on Divine Action*, edited by Robert J. Russell, Kirk Wegter-McNelly, and John Polkinghorne, 111–140. Berkeley: Center for Theology and the Natural Sciences.

Carroll, Sean. 2010. *From Eternity to Here: The Quest for the Ultimate Theory of Time*. New York: Dutton.

Cartwright, Nancy. 1989. *Nature's Capacities and Their Measurement*. Oxford/New York: Clarendon Press/Oxford University Press.

Diacu, Florin, and Philip Holmes. 1996. *Celestial Encounters: The Origins of Chaos and Stability*. Princeton: Princeton University Press.

Earman, John. 1986. *A Primer on Determinism*. Dordrecht: D. Reidel Pub. Co.

Earman, John. 2000. *Hume's Abject Failure: The Argument Against Miracles*. New York: Oxford University Press.

Ellis, George F.R. 1995. "Ordinary and Extraordinary Divine Action." In *Chaos and Complexity: Scientific Perspectives on Divine Action*, edited by Robert J. Russell, 359–395. Berkeley: Center for Theology and the Natural Sciences.

Ellis, George F.R. 2001. "Quantum Theory and the Macroscopic World." In *Quantum Mechanics: Scientific Perspectives on Divine Action*, edited by Robert J. Russell, Kirk Wegter-McNelly, and John Polkinghorne, 259–291. Berkeley: Center for Theology and the Natural Sciences.

Harrison, Peter. 2002. "Voluntarism and Early Modern Science." *History of Science* 40: 63–89.

Larmer, Robert. 2014. "Divine Intervention and the Conservation of Energy: A Reply to Evan Fales." *International Journal for Philosophy of Religion* 75 (1): 27–38.

Leibniz, G.W., and S. Clarke. 1956. *The Leibniz-Clarke Correspondence: Together with Extracts from Newton's Principia and Optics*. Edited by H.G. Alexander. Manchester: Manchester University Press.

Mackie, J.L. 1982. *The Miracle of Theism: Arguments for and Against the Existence of God*. Oxford: Clarendon Press.

Macquarrie, John. 1977. *Principles of Christian Theology*, 2nd ed. New York: Charles Scribner.

Murphy, Nancey. 1995. "Divine Action in the Natural Order." In *Chaos and Complexity: Scientific Perspectives on Divine Action*, edited by Robert J. Russell, Nancey Murphy, and Arthur R. Peacocke, 325–357. Berkeley: Center for Theology and the Natural Sciences.

Murray, Michael J. 2003. "Natural Providence (or Design Trouble)." *Faith and Philosophy* 20 (3): 307–327.

Norton, John. 2008. "The Dome: An Unexpectedly Simple Failure of Determinism." *Philosophy of Science* 75: 786–798.

Omnès, Roland. 1994. *The Interpretation of Quantum Mechanics.* Princeton: Princeton University Press.

Papineau, David. 2009. "The Causal Closure of the Physical and Naturalism." In *The Oxford Handbook of Philosophy of Mind*, edited by B. McLaughlin, A. Beckermann, and S. Walter, 53–65. New York: Oxford University Press.

Peacocke, Arthur. 1993. *Theology for a Scientific Age : Being and Becoming—Natural, Divine and Human*, 2nd ed. London: SCM Press.

Plantinga, Alvin. 2008. "What Is 'Intervention'?" *Theology and Science* 6 (4): 369–401.

Plantinga, Alvin. 2011. *Where the Conflict Really Lies: Science, Religion, and Naturalism.* New York: Oxford University Press.

Polkinghorne, John. 1989. *Science and Providence: God's Interaction with the World.* Boston: New Science Library.

Polkinghorne, John. 1994. *The Faith of a Physicist.* Princeton: Princeton University Press.

Polkinghorne, John. 1995. "The Metaphysics of Divine Action." In *Chaos and Complexity : Scientific Perspectives on Divine Action*, edited by Robert J. Russell, Nancey Murphy, and Arthur R. Peacocke, 147–156. Berkeley: Center for Theology and the Natural Sciences.

Polkinghorne, John. 1996. "Chaos Theory and Divine Action." In *Religion and Science: History, Method, Dialogue*, edited by Mark W. Richardson and Wesley J. Wildman, 243–252. New York: Routledge.

Pollard, William G. 1958. *Chance and Providence: God's Action in a World Governed by Scientific Law.* New York: Scribner.

Ratzsch, Del. 1987. "Nomo(theo)logical Necessity." *Faith and Philosophy* 4 (4): 383–402.

Ruelle, David. 1994. "Where Can One Hope to Profitably Apply the Ideas of Chaos?" *Physics Today* 47 (7): 24–30.

Ruse, Michael. 2001. *Can a Darwinian Be a Christian?: The Relationship Between Science and Religion.* New York: Cambridge University Press.

Ruse, Michael. 2012. "How Not to Solve the Science-Religion Conflict." *The Philosophical Quarterly* 62 (248): 620–625.

Russell, Robert J. 2002. "God's Providence and Quantum Mechanics." *Counterbalance.* http://www.counterbalance.net/physics/qmprovid-frame.html. Accessed April 11, 2013.

Russell, Robert J. 2008a. "Quantum Physics and the Theology of Non-Interventionist Objective Divine Action." In *The Oxford Handbook of Religion and Science*, edited by Philip Clayton, 579–595. New York: Oxford University Press.

Russell, Robert J. 2008b. *Cosmology: From Alpha to Omega*. Minneapolis: Fortress Press.

Saunders, Nicholas. 2000. "Does God Cheat at Dice? Divine Action and Quantum Possibilities." *Zygon* 35 (3): 517–544.

Saunders, Nicholas. 2002. *Divine Action and Modern Science*. Cambridge: Cambridge University Press.

Sober, Elliott. 2011. "Why Methodological Naturalism?" In *Biological Evolution: Facts and Theories. A Critical Appraisal 150 Years After "The Origin of Species"*, edited by G. Auletta, M. Leclerc, and R.A Martínez, 359–378. Roma: Gregorian & Biblical Press.

Stewart, Ian. 1989. *Does God Play Dice?: The Mathematics of Chaos*. Cambridge: Blackwell.

Stoeger, William. 1995. "Describing God's Action." In *Chaos and Complexity: Scientific Perspectives on Divine Action*, edited by Robert J. Russell, Nancey Murphy, and Arthur R. Peacocke, 239–261. Berkeley: Center for Theology and the Natural Sciences.

Tennant, F.R. 1924. "Theism and Laws of Nature." *The Harvard Theological Review* 17 (4): 375–391.

Tracy, Thomas. 1995. "Particular Providence and the God of the Gaps." In *Chaos and Complexity: Scientific Perspectives on Divine Action*, edited by Robert J. Russell, Nancey Murphy, and Arthur R. Peacocke, 289–324. Berkeley: Center for Theology and the Natural Sciences.

Tracy, Thomas. 2003. "Divine Action and Modern Science." *Notre Dame Philosophical Reviews*, October 9. https://ndpr.nd.edu/news/23530-divine-action-and-modern-science/. Accessed April 26, 2013.

Tracy, Thomas. 2008. "Theologies of Divine Action." In *The Oxford Handbook of Religion and Science*, edited by Philip Clayton, 596–611. Oxford: Oxford University Press.

Van Fraassen, Bas C. 1989. *Laws and Symmetry*. Oxford: Clarendon Press.

Wald, Robert M. 2010. *General Relativity*. Chicago: University of Chicago Press.

Ward, Keith. 1990. *Divine Action*. London: Collins.

Wilson, Mark. 1989. "Critical Notice: John Earman's a Primer on Determinism." *Philosophy of Science* 56 (3): 502–532.

5

Naturalisms and Design

5.1   Science, Myths, and Legends

Quick! Think of something that everyone believed a thousand years ago but
was false. My students always come up with the same answer: people in the
Middle Ages believed that the Earth is flat. Many of those same students
were also taught, as I was, that Christopher Columbus sailed west to prove
that the Earth is round.

Both the Columbus story and the ancient-people-flat-Earth claim are
demonstrably false (Cormack 2009). Aristotle knew that we live on a sphere.
So did everyone else in the western medieval world with any education.
Columbus? He was trying to beat the competition to India. The shape of the
Earth was a given.

One of the other myths that we have been chipping away at is the supposed state of war between science and religion. The relation between the
two is complex, but we have not yet seen a fundamental conflict between
the two. Perhaps, one might suggest, we've been looking in the wrong place.
Physics may not have any direct conflicts with religion, but biology has not
been so lucky. Old battles with creationism and new ones with intelligent
design (ID) theory point to conflict, not harmony.

All the press in the last 10 years has certainly generated a lot of controversy over ID. Whether this is itself a conflict between religion and science
is another matter. In my view, most of what one finds on the Internet and in
the news regarding ID has more to do with the culture wars than science.
For our purposes, a distinction can be made between (i) arguments for ID

The Physics of Theism: God, Physics, and the Philosophy of Science,
First Edition. Jeffrey Koperski.
© 2015 John Wiley & Sons, Ltd. Published 2015 by John Wiley & Sons, Ltd.

and (ii) the ID *movement* associated with William Dembski and the Discovery Institute. There is wide support among theists for the notion of design in nature; the ID movement is another matter. Many agree with the Harvard astronomer Owen Gingerich: "I believe in intelligent design, little *i* little *d*" (2005).[1] While most ID advocates are Christian theists, most Christian academics are not in favor of the broader sociopolitical agenda of the Discovery Institute.

A note before we begin. While this chapter goes somewhat beyond the physical sciences, it is a necessary digression. ID raises several challenges for the relation between science and religion. One's views on these matters ramify across the other sciences, including physics. In particular, the relation between design arguments and naturalism transcends biology. So while we could have left ID out of the conversation and simply discussed the boundaries of science, varieties of naturalism, and the mechanics of theory change head on, it seems best to discuss these matters in the context of a live debate.

I leave aside political matters such as whether ID can ever be mentioned in the public schools. Even so, there are still many issues to discuss. Can design, especially supernatural design, play any legitimate role in science? Is the ID question just a matter of evidence? What is the proper role for naturalism in all this? These are important questions in the philosophy of science. Before taking them up, let's briefly look at the core concepts used in ID today.

5.2   Intelligent Design

The best of the scientific arguments for ID in biology come from biochemist Michael Behe. He believes that current evolutionary theory cannot explain the *irreducible complexity* of some biological systems. The basic idea is found in—of all places—Darwin's *The Origin of Species*: "If it could be demonstrated that any complex organ existed which could not possibly have been formed by numerous, successive, slight modifications, my theory would absolutely break down" (1869, 169). Behe claims to have found them:

> A system which meets Darwin's criterion is one which exhibits irreducible complexity. By irreducible complexity I mean a single system which is composed of several interacting parts that contribute to the basic function, and where the removal of any one of the parts causes the system to effectively

cease functioning. An irreducibly complex system cannot be produced gradually by slight, successive modifications of a precursor system, since any precursor to an irreducibly complex system is by definition nonfunctional. (2001, 247)

An irreducibly complex system can only perform its function as a whole. Without each of the parts fitted together just so, the system would be useless. But useless systems are not chosen by natural selection since they do not improve the fitness of an organism. The challenge then is to show how mutation and natural selection could ever produce a system that, until all the parts are present and working together, confers no competitive advantage. Before the whole starts working, there is nothing there for natural selection to select. Useless mechanisms cannot get a foothold in the gene pool via natural selection with the hope that one day they might be useful.

Irreducibly complex systems are at the very least a challenge for textbook evolutionary biology. Behe, of course, offers another explanation: design. Irreducibly complex structures have the look and feel of the "purposefully arranged" objects we normally think of as machines. He doesn't deny that mutation, natural selection, and common ancestry have a part to play in the grand evolutionary scheme. He does deny that these undirected components of evolution are sufficient to explain the sorts of things biochemists have discovered. In one way or another, design, purpose, and teleology must be allowed to return from scientific exile.

Mathematician William Dembski takes a more general approach that is not limited to evolution. Biologists have long recognized that the distinction between software and hardware has analogies in biological systems. DNA contains information that is distinct from the medium that encodes it. Therefore, two distinct explanations are needed: (1) how the material medium came to be and (2) how this material came to be encoded with the specific information it contains. Dembski argues that systems displaying *complex specificity* cannot be explained naturalistically.[2] He begins with the Shannon information theory, which electrical engineers typically use to design communications systems. Dembski borrows the idea that the greater the number of possible messages encoded, the greater the information in a given message. Say Smith receives the coded message *alpha gamma*. How much information does it contain? It depends. What is the space of all possible messages? Perhaps, there are just two possibilities: <*alpha, gamma* > and < *gamma, alpha*>. Or perhaps, there are six: <*alpha, gamma*>,

<beta, gamma>, *<delta, gamma>*, and so on. Smith's message would contain more information in the latter case since the message has been selected from a larger space. So when Dembski talks about *complex* information, he means a very high amount of information—a sequence selected from a large space of possibilities. There is a very low probability of randomly selecting any particular sequence from this space, just as there is a very low probability of getting a particular string of numbers from a thousand rolls of a pair of dice.

Specificity is a trickier notion, but intuitively, the idea is that the information fits a pattern. Consider the extraterrestrial signal heard by Jodi Foster's character in the movie *Contact*. A long string of pulses fits a recognizable pattern: the prime numbers between 1 and 100. Since these numbers were chosen from among all the positive integers, the string of pulses was also complex (i.e., low probability of occurring randomly). And since the string came from space, she makes the correct inference: these pulses are a sign of nonhuman intelligence. Nature cannot produce a long, improbable sequence that conforms to a specified pattern. After all, if the leaves on your lawn did happen to spell out OHIO, what would you conclude?

From there, Dembski's argument is quite simple. All known examples of complex specificity have been produced by intelligent beings. Nature has never created nor significantly increased the complex specified information of any organism. Nature can produce random sequences, like the path of a blowing leaf. Nature can also produce regular sequences, like the orbit of a planet. But no combination of chance and natural law, Dembski argues, can create complex specified information. Hence, we should take that property as a reliable indicator of design.

Dembski and Behe both conclude that nature contains signs of non-human intelligence. Although the God of classical theism is not the only possible candidate, that is certainly where most ID advocates are placing their bets. The reason most ID proponents are Christian is due to the hope that design theory might provide indirect evidence of a creator. (Ironically, many young earth creationists originally opposed ID because of this limited conclusion. If ID could not prove the existence of the God of the Bible, they believed that it was little better than theistic evolution.)

There is a lot one could say about Dembski and Behe's arguments, but that would take another book.[3] For our purposes, this introduction is enough to show what the critics are taking aim at.

5.3   It's Not Science

Even if science and religion aren't in a state of war, they usually deal with very different subjects. If the designer in intelligent design is in fact God, then it would seem that ID is a theological matter, not a scientific one. ID advocates, on the other hand, insist that the arguments and evidence for their views are purely scientific. One issue, then, is whether ID counts as science. There are two main reasons for thinking not.

5.3.1   Demarcation Part 1: Motives

One of the stated goals of ID is to push biology out of a cul-de-sac, the degenerating theory of neo-Darwinism. But what is it *really* about? Is it, as Forrest and Gross put it in the title of their book, "Creationism's Trojan Horse" (2004), a way of sneaking good old-fashioned creation science past the censors? ID critics think so. They prefer the label 'intelligent design creationism' to help drive the point home. ('Neocreos' and even 'country-bumpkin creos' may be used when one is feeling less charitable.) Even federal judges agree. *Kitzmiller v. Dover* was a 2005 lawsuit in which the Dover, Pennsylvania, School Board was forbidden from requiring a pro-ID disclaimer to be read in biology classes when evolution was discussed. Judge Jones agreed that the "evidence at trial demonstrates that ID is nothing less than the progeny of creationism" (*Kitzmiller v. Dover* 2005, 40).

Regardless of what labels we use, motivation appears to be a key question in this debate. Fortunately, an "anonymous source" at the Discovery Institute tells us what ID proponents are really after, reports Forrest, in "an internal... document, titled 'The Wedge Strategy,' " (2001, 3). Judge Jones was quite impressed with this information in *Kitzmiller*, citing it as decisive evidence that ID plans to replace current science with "Christian science" (2005, 36). To further this goal, says Forrest, "the [Discovery Institute] creationists have taken the time and trouble to acquire legitimate degrees, providing them cover both while they are students and after they join university faculties" so as to "blend more smoothly into the academic population" (2001, 38–39). Here, then, is proof that the Discovery Institute, whose fellows are the leaders of ID, has religious motivations. Even if they are not traditional creationists, they are predominately Christian, and their hope is that ID will lead to a revolution that will overthrow naturalistic science.

Many take this as proof that ID is therefore fundamentally a religious notion, not a scientific one.

5.3.2   *Demarcation Part 2: Methodological Naturalism*

A more prevalent view among academics is that ID is not science because it fails to be properly naturalistic. Science, they say, can only study natural entities. While one might argue that this begs the question against ID, critics like Robert Pennock reply that a distinction must be made between methodological and metaphysical naturalisms (Pennock 1996). The latter is the view that nothing exists beyond the natural causal order. It is the descendant of what Enlightenment thinkers called 'materialism': everything that exists is made of matter. Methodological naturalism (MN), on the other hand, has to do with theory formation: researchers must proceed *as if* metaphysical naturalism were true regardless of what they believe about anything else. Immaterial entities such as spirits, souls, and final causes have long been rejected in science, and this rejection is neatly captured by MN. (Of course, both types of naturalism could be abbreviated MN, but I'll only use it for methodological naturalism as it is so frequently mentioned in this chapter.)

It is therefore not the case, Pennock argues, that neo-Darwinists dogmatically refuse to consider ID because of some antisupernatural bias. Rather, since ID posits a nonphysical intelligence, it violates MN and is therefore not science.[4] Strictly speaking, this is false, since ID does not entail theism. ID theorists readily admit that the evidence only points to an intelligence of some kind or other. It does not narrowly confirm the existence of God. Let's ignore that for now. People on both sides of the debate commonly assume that ID violates MN.

Why is MN so important? One reason given is that if God were to tinker with nature, we could not trust the laws of nature. "Without the constraint of lawful regularity, inductive evidential inference cannot get off the ground" (Pennock 2001, 88). Hence, we must presuppose that God does not interfere with natural law in order to do science. A closely related reason is that design explanations would hamper the progress of science. Since "God did it" is potentially an answer to any why question, allowing design back into biology is a science stopper:

> Why bother to conduct an exhaustive molecular search through simian virus genomes to find the source of HIV when clear-thinking ID scholars have

concluded that it was sent as a divine warning against deviant lifestyles?… A theistic science may be friendly to the tenets of faith… but it will no longer be the science we have known. It will cease to explore, because it already knows the answers. (Miller 2009, 197–198)

If divine fiat were an acceptable explanation, why push on with difficult and expensive research? Many ID critics argue that MN is essential for motivating scientific progress. We will examine these arguments more closely in Section 5.4.2.

5.3.3   Theists for MN

One can understand why a metaphysical naturalist would favor MN. If there aren't any beings beyond the natural order, then science isn't missing anything by ruling them out. It is somewhat surprising, though, that many theists also support MN.

Although the phrase 'methodological naturalism' is recent, the idea arose in the seventeenth century as a doctrine akin to separation of church and state. Those in the humanities did not want scientists getting involved in matters outside of their expertise, and natural philosophers did not want clergy and kings looking over their shoulders. In 1660, Charles II set the parameters for the Royal Society of London to be, in the words of Robert Hooke, "to improve the knowledge of natural things, and all useful Arts, Manufactures, Mechanics, Practices, Engynes and Inventions by Experiments (not meddling with Divinity, Metaphysics, Moralls, Politicks, Grammar, Rhetoric, or Logick)" (Proctor 1991, 33). By whatever name, something like MN was useful in helping science form into its own set of specialized disciplines, distinct from philosophy and theology.[5]

Beyond the history of science, there are three reasons theists, especially those in academia, embrace MN. The first has to do with divine action. As we saw in the previous chapter, MN is completely reasonable assuming that noninterventionism is correct. If God, as a matter of principle, doesn't make changes to the natural order, then scientists won't miss anything by ignoring supernatural causes. This is a large part of Michael Murray's argument against ID (2003).

Second, each science requires training for research within a narrow domain. Scientists are not equipped to deal with knowledge beyond their sphere. As Ratzsch puts it, "nonnatural things operate in ways beyond the grip of any empirical method… consequently, such concepts could operate in ways beyond any reliable (i.e., empirical) methodological controls, checks, or constraints" (Ratzsch 2001, 96). Metaphysics, theology, and ethics are therefore beyond the range of the scientist's professional competency. MN keeps scientists from straying into areas beyond their training and then calling it 'science.'

This is closely related to the third reason some theists support MN: not all knowledge can be reduced to scientific data and theories. Even many metaphysical naturalists believe there are truths captured by ethics, history, and philosophy that science cannot reach. For example, boiling kittens for entertainment is wrong, but that ethical truth cannot be derived from the natural sciences. Theists would add theological truths to this list. MN is just a way of keeping scientists from treading on areas of knowledge beyond their expertise. Naturalistic boundaries help keep scientists, including those in the social sciences, from trying to reduce theological truths to scientific ones.

As we've seen, a multipronged case can be made for MN from both a theistic and naturalistic point of view. Some philosophers of science have nonetheless turned a critical eye toward MN in recent years. Let's now consider the arguments against.

5.4   Faulty Demarcation

5.4.1   Motives Don't Matter

Let's put aside the question of demarcation for a moment and focus on the rhetoric. Is equating creationism with ID accurate? 'Creationism'—much like 'evolution'—is rather ambiguous. As one of the members of our biology department defines it, all theists are creationists. But that can't be right. Kenneth Miller is a well-known ID critic and a Roman Catholic. No one familiar with the debate would consider him a creationist. The term is properly used for those who believe in a literal reading of Genesis, which entails that the Earth is less than 20 000 years old (Sober 2011, 360–361). If you were to bump into a person on the street who supports ID, that person would likely be a traditional creationist of this sort. Among prominent ID leaders, however, philosopher Paul Nelson is the only classical creationist as far as I know. Others, like biochemist Michael Behe, accept the common ancestry of species and have no particular qualms about the fossil record. In other words, Behe believes in macroevolution; it's the sufficiency of the Darwinian mechanism that he doubts.

In my view, labeling those who doubt the efficacy of genetic mutation and natural selection "creationists" is a rhetorical strategy, what one logic text calls *stereotyping*. Cable television provides ready exemplars for both the creationist stereotype and its cousin, the fundamentalist.[6] Critics try to shape the debate by connecting ID to these templates. If successful, little work needs to be done. The labels tell us who represents the side of rationality over and against the side of ignorance. Having sorted "us" and "them," what "they" actually say matters little, whoever "they" happen to be. We must recognize that while this is a common argumentative strategy in talk radio and national politics, it is not itself a logical critique. Placing the black hat on one's opponent is no substitute for an argument.

Getting back to the religious motivations of ID theorists, an important question remains: So what? How is this information relevant to the rational assessment of ID? Consider an analogy. When I was a graduate student, one of my professors was a committed Marxist. As the faculty advisor for a socialist student group, he made it clear that he had wanted to become a professor in order to promote his political views. He hoped to persuade students accordingly. Now, consider the articles he had published in scholarly journals. Did the fact that he had a political motivation affect the strength of his arguments in those papers? Should the editors of those journals have taken his political agenda into consideration in deciding whether to publish them?

As every logic student knows, the answer is "no." One's motivations for presenting an argument have no bearing whatsoever on the strength of that argument. Evaluating a conclusion by questioning the motives of its source is an *ad hominem* attack. Arguments must be judged on their merits regardless of who presents them or why. Lysenko's theory of inheritance wasn't bad because the Communist Party in the Soviet Union promoted it; it was bad because his theory was an experimental failure. In contrast, Martin Luther King, Jr., was certainly motivated by his religious beliefs, as was William Wilberforce. That psychological fact neither adds to nor detracts from the strength of their arguments. Stereotyping, *ad hominem*, and appeals to anger are effective rhetorical devices, but they all interfere with rational investigation. Consider this passage where Sober catches Pennock making just such a move:

> Notice the shift from propositions to people in the passage I quoted from Pennock. He begins by discussing a *proposition*… and then shifts to a fact about how creationists defend this proposition, pointing out that "*creationists*

have a fundamentally different notion from science of what constitutes proper evidential grounds for warranted belief" [emphasis Sober]. It is true that creationists have been unscientific, but this is a fact about them; nothing follows about the character of the theory they wish to defend. Consider a dogmatic Darwinian or a dogmatic Newtonian who argues unscientifically; this fact about them does not show that their pet theories are unscientific. (Sober 2011, 370)

Sober highlights the difference between an argument and its source. Logical critiques must be aimed at the former, not the latter. Once a person has presented an argument—premises and conclusion—counterarguments aimed at that person's motives are textbook fallacies.

Textbook or not, philosopher Christopher Pynes (2012) has recently argued that ad hominem arguments are legitimate in this context. He raises an interesting point. If a person is providing testimony, then motives matter in the weighing of his or her claims. You *should* consider the salesperson's motivation when presenting the facts about his or her product. The police should be skeptical when questioning suspects with mob connections. Likewise, says Pynes, one should question the religious motives of ID theorists.

While I agree with the first two examples, they are poor analogies for ID. Behe, for example, is not offering *testimony* about events in his lab; he is offering *arguments* for design based on biochemical systems—his area of expertise. Criticizing Behe's arguments by attacking his possible religious motivations is a paradigm case of an ad hominem fallacy. This all becomes clear if we turn our attention to Forrest. Is it legitimate to answer her objections to ID by questioning her motives? Forrest is on the board of directors for the New Orleans Secular Humanist Association (http://nosha.secularhunanism.net). This group actively tries to prove that religious beliefs are based on ignorance and superstition. They strategically promote secular humanism; they hold conferences; they have their own newsletters and publications; they take donations. I submit that Forrest's academic publications are motivated by her antireligious views.

Does it matter? No, not as a matter of logic. Her psychological state is completely irrelevant to the strength of her arguments. In order to rebut a person's arguments, one must directly address those arguments. Anyone trying to undermine Forrest's conclusions by pointing to her antireligious motivations is committing a logical fallacy. The same goes for ID.

But what about the demarcation question? The reason Forrest spends so much time documenting the motives of ID proponents is to show that it's all ultimately about promoting Christianity. ID isn't really a scientific research program at its core. Say we caught a scientist saying something like this off the record:

> When I wrote my book, I had an eye on principles that might help people considering belief in God. Nothing would make me happier than to see it used for that purpose.

According to many ID critics, a person having these kinds of apologetic motives might produce something vaguely scientific, but strictly speaking, it can't be science.

If that principle is correct, then theologians and physicists will be surprised to find that Newton's *Principia Mathematica* should be moved to a different part of the library since, it seems, it isn't science. Here is the original quote:

> When I wrote my treatise about our Systeme… I had an eye upon such Principles as might work with considering men for the beliefe of a Deity & nothing can rejoyce me more than to find it usefull for that purpose. (Davis 1996, 78)

Newton expressed this sentiment in several places. In a draft of the *Principia*'s General Scholium, he claimed that the "dominion or Deity of God is best demonstrated not from abstract ideas but from phenomena, by their final causes." Newton had a religious motive for this work. The same goes for Boyle, Faraday, and many other scientists past and present. This fact does not make their work nonscience or bad science. Good science can be produced from a variety of motivations, including religious ones.

5.4.2 Demarcation and MN

A more frequent objection to ID appeals to MN. Science must be naturalistic. As Judge Jones put it, "ID's failure to meet the ground rules of science is sufficient for the Court to conclude that it is not science" (*Kitzmiller v. Dover* 2005, 91). Scientists, teachers, and textbook writers therefore need not consider it.

That we need MN to rescue science from supernatural design is doubtful. As we'll see, all of the work supposedly done by MN is in fact already done by other metatheoretic shaping principles. Nonetheless, let's grant for the moment MN's status as shaping principle. The problem has to do with the wielding of MN to define ID as religion.

A crucial assumption is that once a concept achieves the status of shaping principle, it becomes an immutable axiom for all future science. That is false, if the history of science is any sort of guide. Almost everything in science has been subject to change from data and models to theories and laws. Like mutual funds, past success does not guarantee future performance. Shaping principles are no exception. In many cases, one desideratum is traded off against another. Consider simplicity. Among competing explanations, we tend to prefer the simple and elegant over the complex and convoluted. But scientists, like the rest of us, routinely ignore this preference due to an implicit *ceteris paribus* (all things being equal) condition. The Standard Model of particle physics has tremendous explanatory power, uniting the strong, weak, and electromagnetic forces. It is also far more complex than anything atomic theorists had envisioned at the turn of the last century. Renormalization methods used to manage its sometimes inconvenient mathematics are neither simple nor elegant. They do, however, work quite well. The point is that, like moral duties, the explanatory virtues sometimes conflict. One shaping principle must sometimes be traded off against another.

Moreover, as we saw in Chapter 1, shaping principles have been suspended and changed throughout the history of science (Section 1.3.2). Aristotelian principles were replaced by Cartesian ones. Cartesian principles did not survive the Newtonian revolution.[7] This sort of flexibility is a necessary condition for advancement. Einstein believed that the universe was static on a large scale. When his field equations showed that space must either expand or contract, Einstein introduced his infamous cosmological constant to allow for a static solution. He soon changed his mind. What if the metaphysical principles behind the static universe had been fixed and were unrevisable? Then Big Bang cosmology would have been ruled out as pseudoscience. Shaping principles cannot be set in stone if science is to adapt to new information.

ID critics have argued that MN, in contrast, is sacrosanct. In his Dover testimony, Pennock claimed that

> This self-imposed convention of science, which limits inquiry to testable,
> natural explanations about the natural world, is referred to by philosophers

as "methodological naturalism" and is sometimes known as the scientific method. (*Kitzmiller v. Dover* 2005, 83)

Naturalistic explanations are not merely desirable; they are "an essential attribute to science by definition and by convention" (84). ID violates MN and thus "by definition," it cannot be science.

All this talk of whether ID is science or religion ignores a long-standing problem in the philosophy of science. It begins with a surprising fact: there is no such thing as *the* scientific method. Philosophers generally agree with Lee Smolin on this: "I'm convinced, like many practicing scientists, that we follow no single method…." (2006, 297). The failed search for a method unique to science is part of what philosophers call the *demarcation problem*. There are no criteria that count every scientific specialization as "in," while other disciplines and pseudoscience remain "out."[8] On my view, the reason for this failure is that there is no strict boundary between, say, science and metaphysics. The two overlap a great deal. Consider this question: Was Bohr's conflict with Einstein (Section 1.3.2) a matter of science or of philosophy? I don't see how one could confidently answer one way or the other.

Neither Pennock nor anyone else has solved the demarcation problem. MN is not an immutable principle that can neatly separate science from nonscience. As philosopher of science Philip Kitcher put it in an anticreationist text, "postulating an unobserved Creator need be no more unscientific than postulating unobserved particles" (1983, 125). W.V.O. Quine expressed a similar view both in his early work ([1951] 1980, 45) and in one of his last articles: "If I saw indirect explanatory benefit in positing sensibilia, possibilia, spirits, a Creator, I would joyfully accord them scientific status too, on a par with such avowedly scientific posits as quarks and black holes" (1995, 252).

The bottom line is this. The future use or suspension of MN depends on what is discovered. If the best explanation for some new phenomenon is design, even supernatural design, that would not bar it from being a scientific explanation. It borders on academic incompetence to pretend that *science* has strict boundaries and then gerrymander those boundaries to keep out the riffraff. Philosophers of science in particular should know better.

But what of the science-stopper objection? Won't the appeal to supernatural causes bring an end to naturalistic research, as biologist Ken Miller testified (*Kitzmiller v. Dover* 2005, 66)? I think this is a plausible objection. If one already has a supernatural answer to a question, why search for another?

Fortunately, we need not rest on intuition to answer this question. The history of science provides ample cases to test the science-stopper claim. Is it the case that design explanations proved to be overly tempting for theistic scientists and so blocked the development of rival theories? ID critics often assume that this *must* have happened sometime or other, hence the need for MN. The history of science remains uncooperative on this point, however.[9] James Clerk Maxwell presented a design argument in the ninth edition of the *Encyclopædia Britannica* in his entry on the atom. I see no evidence that his work in statistical mechanics or electromagnetism was impeded by the possibility of design. As far as I can tell, "God did it" is not an answer that credentialed theistic scientists reach for whenever research bogs down.

To sum up this section, even if MN is a shaping principle within contemporary biology, that would not mean it is an inviolable maxim that scientists must employ come what may. Naturalistic critics of ID may certainly bet that no forthcoming discovery will require design, but they cannot guarantee it. In order to know one way or the other, the evidence and arguments have to be evaluated. And in order to evaluate design claims, scientists must be allowed to consider them *qua* scientists rather than being told that such inquiries must be left to theologians and fundamentalist preachers.

5.4.3   Theism and MN

Metaphysical naturalists are not the only ones in favor of MN, recall. Some theists use MN as a way to keep scientists within their areas of expertise and away from facile reductions of theological truths to naturalistic ones. Again, I have some sympathy for these arguments. Each of the sciences is highly specialized and fragmented. It is difficult for scientists to keep up in their own field, let alone matters in another discipline. There are good reasons for wanting scientists not to tread on areas in which they have no expertise. Nonetheless, there are three reasons why theists should be wary of MN.

5.4.3.1   There Don't Seem to Be Any Real Boundaries to Science
If MN is supposed to be a no trespassing sign, scientists don't take it as such:

> [The] scope of science has always expanded, steadily replacing supernatural explanations with scientific ones. Science will continue this inexorable march…. After all, there is no evidence that consciousness and mind arise from anything other than the workings of the physical brain, and so those phenomena are well within the scope of scientific investigation. What's more,

because the powerful appeal of religion comes precisely from its claims that the deity intervenes in the physical world, in response to prayers and such, religious claims, too, fall well within the domain of science. (Singham 2010)

Scientists just don't seem to have any qualms about pontificating on matters of philosophy and theology. And while the conflict model might be dead among religion scholars, it still lives in the popular psyche and among many scientists:

> The reason why science and religion are actually incompatible is that, in the real world, they reach incompatible conclusions…. Different religions … make very different claims, but they typically end up saying things like "God made the universe in six days" or "Jesus died and was resurrected" or "Moses parted the red sea" or "dead souls are reincarnated in accordance with their karmic burden." And science says: none of that is true. So there you go, incompatibility. (Carroll 2009)

But wait. If science subscribes to MN and MN restricts scientists to their domains of expertise, how could science entail anything about divine action or hermeneutics? The answer is that, while Carroll's examples are paradigmatically supernatural, scientists do not in fact honor the supposed boundaries marked out by MN. MN is only used as a stick with which to beat ID.

Some will reply that these examples are of scientists failing to recognize the boundaries of science. Perhaps so, but it does appear that the boundary only works one way. Scientists can cross at will; those on the religion side must remain where they are.

5.4.3.2   MN Limits the Explanatory Resources of Science

Given the limitations of MN, science is sometimes pushed into odd corners. Recall the discovery of fine-tuning from Chapter 2. As we saw, most agree that this requires an explanation. Under MN, when physicist Lawrence Krauss explains fine-tuning by positing a vast multiverse of possible universes each with different values for these constants, he's doing science. When astronomer Owen Gingerich explains the very same observations by means of design, he's doing religion. This, it seems to me, is completely artificial and ad hoc. As the SETI project and archeology show, design is an explanatory concept already used in science. Nonetheless, the best scientifically acceptable explanation of fine-tuning under MN is an undetectable multiverse. Cosmologists must therefore pursue that hypothesis to win grants, publish papers, and get tenure, even if supernatural design happens to be the right answer.

That is a rather odd result. Almost everyone agrees that God *might* exist even if there isn't sufficient evidence to believe it. If so, then God might have literally fine-tuned the universe in order for life to exist. As a thought experiment, let's just stipulate that this has happened. Let's say God exists and directly fixed the values of the fine-tuned cosmic constants. If science must be naturalistic, then since (i) the fine-tuning data cries out for an explanation and (ii) the only explanations allowed are naturalistic, science would be driven into accepting false explanations. MN is therefore potentially in conflict with realism. In order to hold scientific realism, one must believe that mature theories are generally reliable indicators of truth. But if there is a choice between naturalism and truth, MN forces science to choose the former. Once science is limited to certain kinds of entities, it can no longer follow the data wherever it leads. Science is instead forced to beat the data until it offers a naturalistic confession.

Another problem for MN is that no one knows what sort of explanatory resources science will need in the future. One can bet that we will never need to use design, but that's a prediction, not a discovery or an inference from established truths. Many of the expectations of late nineteenth-century physicists were dashed by general relativity, quantum mechanics, and chaos theory. How can one guarantee today which ideas will and will not be needed in the next century?

5.4.3.3   *MN Is Superfluous*

Science no doubt became increasingly naturalistic in the eighteenth and nineteenth centuries, eschewing explanations that appealed to God's direct intervention in favor of natural causes.[10] In a much less secular age, MN had some utility in helping natural philosophy evolve into specialized sciences. That's not to say that MN was always needed. Many historical confrontations between naturalistic and design hypotheses were settled by appeal to simplicity in the form of Ockham's razor. To cite one (overly used) example, when Laplace presented Napoleon with a copy of his *Mécanique Céleste*, the emperor wished to know why it did not contain any reference to God. Laplace replied, "I have no need of that hypothesis" (Herschel 2010). Nor did he need to invoke MN in order to make his case.

What about today? Is MN needed to rescue science from supernatural design here in the secular west? In my view, it serves no useful purpose. If one closely examines the arguments, MN is almost always a placeholder for some other shaping principle. ID critics often begin by invoking MN but then seamlessly switch to complaints about testability and fruitfulness. Biologist Ken Miller provides one such example:

> Supernatural design is a comforting message, but beyond it, ID has almost nothing else to say. It cannot tell us why our bodies are "designed" the way they are, and it has no explanation for the patterns of the fossil record or our similarities to other organisms—except to claim that that was simply the way the designer chose to make it. (2009, 219–220)

Pennock similarly defends MN as a necessary condition for empirical confirmation:

> Science operates by empirical principles of observational testing; hypotheses must be confirmed or disconfirmed by reference to empirical data…. Science assumes Methodological Naturalism because to do so otherwise would be to abandon its empirical evidential touchstone. (2001, 89)

There is a legitimate worry here. Design claims, especially supernatural ones, are inherently difficult to test, and it's hard to know what to do with such a claim in terms of ongoing research. Theories must, in Ratzsch's terms, somehow "be put in *empirical harm's way*" (2001, 98).[11] Instead, design inferences tend to be hyperflexible, escaping "any attempt of nature to nail it with refuting data, happily absorbing the data in question" (Ratzsch 2001, 112).[12] Here's an example of what Ratzsch has in mind. Say there are phenomena that indicate design based on Behe's *irreducibly complexity* or Dembski's *complex specificity*. What if we eventually find good naturalistic explanations for each of these cases? Design theorists could always then demand an explanation for why the laws of nature produce design-like entities—not the first time such a move has been made in the history of design arguments (Numbers 2008, 276). Instead of design being refuted, it simply gets pushed back to the level of laws. This allows ID advocates to claim victory if naturalistic explanations cannot be found but then to shift to new ground when they are. Theories that are easily preserved come what may are hyperflexible—a kind of negative shaping principle, something to be avoided. We prefer concrete theories that might possibly be undermined by contrary data.

My point is this: fruitfulness, testability, and concreteness *are* well-established shaping principles that can be defended in their own right, but none is equivalent to MN. MN could be dropped without loss since the work it supposedly does is carried out by other shaping principles. There is no need to impose MN as an *a priori* restriction. There are other shaping principles in place that ID will have to contend with, just like any other hypothesis, model, or theory. Unlike MN, these other principles are used throughout the sciences rather than merely to criticize one particular foe. For various reasons then, theists should follow Quine—himself an ardent naturalist: if the best explanation for some physical phenomenon is design, even supernatural design, that would still count as a scientific explanation.

At the end of the day, ID turns out to be scientific after all. That is only a small victory, however. The strongest objections have not yet been raised.

5.5   The Real Problems

So why is ID not yet out of the woods? First, to recognize a proposal as scientific does not entail that it is good science. Second, while a given hypothesis might answer an important question, it might also be far more radical than is needed. Both of these are problems for ID.

5.5.1   Good Science

Although ID is not merely creation science under new management, the older debate between Darwinism and creationism is instructive. In my view, Larry Laudan got it exactly right:

> Rather than taking on the creationists obliquely and in wholesale fashion by suggesting that what they are doing is "unscientific" *tout court* (which is doubly silly because few authors can even agree on what makes an activity scientific), we should confront their claims directly and in piecemeal fashion by asking what evidence and arguments can be marshalled for and against each of them. The core issue is not whether Creationism satisfies some undemanding and highly controversial definitions of what is scientific; the real question is whether the existing evidence provides stronger arguments for evolutionary theory than for Creationism…. Debating the scientific status of Creationism… is a red herring that diverts attention away from the issues that should concern us. (1982, 18)

Replace 'creationism' with 'ID' and Laudan could resubmit this as a response to the Dover decision. ID must be judged on its merits.

So then, how does it stack up? If ID is indeed scientific, is it good science? As even its most staunch supporters are willing to admit, that is a much more difficult case to make. Although ID has the attention of researchers on both sides, design-driven science has failed to keep up with the publicity. The bulk of ID science falls into two categories. The first is a host of biological systems that are problematic for neo-Darwinism, usually ones that display irreducible complexity in Behe's sense. ID critics have replied by showing that although complex systems like the bacterial flagellum are improbable, they are still consistent with neo-Darwinism. In other words, there are many soft anomalies in the literature but no hard ones. A hard anomaly is an observation that *cannot* be explained in terms of the reigning theory. For example, black body radiation and the photoelectric effect showed there must be something wrong with classical models of the atom. Some change or other had to be made in order to accommodate the new observations. In contrast, no changes are required in order to accommodate soft anomalies. While they might not comfortably fit within the reigning theory, soft anomalies are strictly speaking possible according to that theory. It may be difficult to imagine how a combination of mutation and natural selection could produce highly integrated systems like the bacterial flagellum. Complex systems are nonetheless logically consistent with neo-Darwinism. ID has yet to produce a hard anomaly.

The second sort of ID-science currently available is research that fits nicely within a design framework, even though the researchers themselves do not support ID. For example, ID proponents often point to publications showing that the so-called "junk DNA" is actually functional. While DNA contains the genetic code for producing protein sequences in every cell, large chunks of DNA appear to be functionally useless. ID proponents have long predicted that research would eventually find some function for non-coding DNA, believing that there is more purpose in biological structures than would be expected from a Darwinian point of view. The nanotechnology approach to microscopic systems is also considered very ID-friendly (Gazit 2007). The reason scientists find such utility in thinking of biological systems as machines is because, in some sense, that's what they are. The conceptual link with human artifacts is not merely a metaphor.

These kinds of ID-related science are well and good as far as they go. What critics rightly demand, however, is peer-reviewed research in which design has more than a mere heuristic role. To be fair, there are more published papers out there than most people realize.[13] And, as ID proponents argue, there is a strong bias against design-motivated articles getting into academic journals.[14] Editors will not risk giving aid and comfort to the enemy. In my view, the ID community is itself partly to blame for this. Some think of ID primarily as a weapon in the culture wars. Antidesign bias in the academy is part of the backlash. Had ID consistently emphasized research over public exposure, the atmosphere of the debate would be different today. Instead, Phillip Johnson and others believed that their ideas were so compelling that once disseminated, ID thought would sweep across the land. A 2001 front-page story in *New York Times* was a cause for much celebration not because it was pro-ID—it wasn't—but because it helped place the debate in the public eye. This has proved to be a failed strategy.

Tactics aside, is ID a thriving research program by any measure? No, not in my opinion. Bias is an obstacle and it is hard to say how many would come out of the closet if there were no risk to their careers. By my estimation, there are hundreds of scientists who would like to make a contribution. They believe there is something critically flawed about the neo-Darwinian paradigm and that design is a better explanation for what they observe. Yet they withhold their professional opinions because of what it would mean for their careers in the present climate.

With that said, bias is not the biggest obstacle for ID. The elephant in the ID room is the lack of a clear vision of what design research might be. Although Dembski has some broad suggestions (2004a, 310–317), the average design-friendly scientist still doesn't quite know what to do. The reason for this traces back to Chapter 1 (Section 1.2). As the late-medieval voluntarists argued, God could have created the universe in many different ways. With so many options, it is hard to say *a priori* how God might have proceeded. Our only recourse is to go out and investigate. The voluntarists understood that reason alone does not get you very far from the premise that God designed nature. There is no way to know precisely how the Designer might have gone about his business. Turning out concrete predictions will therefore be difficult and perhaps impossible. Hence, even if ID is exactly right, there doesn't seem to be much one can do with it. As such, it is hard to see how it can be the basis of a thriving, fruitful research program.

5.5.2   *More Radical than Necessary*

There is a less visible but equally menacing problem for the viability of ID. MN gets a lot of attention, but there is another methodological shaping principle to be contended with, namely, *conservatism*. When faced with anomalous new data, scientists prefer incremental change to more revolutionary change. A second way one may properly criticize ID is to show that it is a more radical proposal than is needed in order to accommodate the evidence against neo-Darwinism.

There are two related but distinct ideas here. One is *epistemic conservatism*, which normally refers to a king-of-the-hill approach: one should keep one's current set of beliefs unless something better comes along to displace them. The fact that there are other possible views, even equally good ones, is not itself a reason to reject one's justified beliefs.

A second version may be traced to Quine's doctrine of *minimal mutilation*: new observations may force a change in one's beliefs, but one should make the smallest change possible in order to accommodate the new information ([1951] 1980, 42–44). What Quine actually said in "Two Dogmas of Empiricism" was that minimal mutilation is "our natural tendency"—a descriptive claim. In the hands of philosophers of science like Larry Sklar (1975), minimal mutilation becomes normative. Belief change should be minimal; dramatic changes ought to be considered only when necessary. This is now a widely accepted view among epistemologists (Lycan 1988, 157–177).

It is a short step, as Quine himself suggested, from this doctrine about one's own beliefs to those of the scientific community. If scientists practice minimal mutilation individually, then their theories will also tend to develop in a conservative way. As new discoveries are made, the body of scientific knowledge should change as little as needed in order to accommodate them. I will refer to this normative principle of theory change as *scientific conservatism*.

To see how this applies to ID, we must recognize that there is a legitimate, scientific controversy over the mechanisms of macroevolution. ID critics often downplay these debates so as not to benefit the enemy, but they are there nonetheless. (In fairness, ID advocates often exaggerate the controversy.) Molecular biologist James Shapiro describes the situation this way:

> [The] debate about evolution continues to assume the quality of an abstract and philosophical "dialogue of the deaf" between Creationists and Darwinists. Although our knowledge of the molecular details of biological organization is undergoing a revolutionary expansion, open-minded discussions of the impact of these discoveries are all too rare. The possibility of a non-Darwinian, scientific theory of evolution is virtually never considered.... I propose to sketch some developments in contemporary life science that suggest shortcomings in orthodox evolutionary theory and open the

door to very different ways of formulating questions about the evolutionary process. (Shapiro 1997)

Paleontologist Stephen J. Gould likewise rocked the boat of (what he later called) "Darwinian fundamentalism" with his theory of *punctuated equilibrium*, but his more damaging proposal had to do with macromutations—large-scale, systemic mutations in a single generation. This was not quite the "hopeful monster thesis" that prompted such outrage against geneticist Richard Goldschmidt in the 1940s, but neither was it the accumulation of tiny changes required by Darwin. Evolutionary biologists remain unimpressed with Gould and their mutual dismissal continued to the end of his life. The polemical baton has been picked up by Simon Conway Morris, who has little respect for "ultra-Darwinists" and

> their almost unbelievable self-assurance, their breezy self-confidence.... [Far] more serious, are particular examples of a sophistry and sleight of hand in the misuse of metaphor, and more importantly a distortion of metaphysics in support of an evolutionary programme. (2003, 314)

Other proposals that are neither design based nor fully Darwinian come from Stuart Kauffman (self-organization and autocatalysis), Brian Goodwin (morphogenesis and developmental constraints), Shapiro (natural genetic engineering), Conway Morris (phenotype convergence), and Sean B. Carroll *et al.* (evo-devo).[15] This list is far from complete.

This would seem to be good news for ID. Many biologists acknowledge that neo-Darwinism has serious anomalies. It is not the case, therefore, that all Darwinian critics fit *Inherit the Wind* stereotypes.

My view is different. In light of scientific conservatism, these non-Darwinian and quasi-Darwinian proposals in fact undermine the viability of ID. Here's why. Even if orthodox neo-Darwinism collapses, design is not the only alternative. More importantly, the rivals are more conservative vis-à-vis the reigning theory. Very little would have to be added to textbook evolutionary theory if one or more of these are accepted. If any one of them is capable of resolving the problems posed by complex structures and macroevolution, then ID is a more radical solution than is needed. Scientific conservatism will thereby continue to undermine the acceptability of ID.

At the end of the day, a host of shaping principles stands against ID even if MN were discarded. Conservatism and fruitfulness get less attention, but they allow the critics to rightly contend that while ID may be scientific, it remains fringe science. These two shaping principles will also be potential hurdles for any new design argument, related to biology or not. If the data in question already have an established scientific explanation, then any new, design-based explanation will face the headwinds of conservatism and questions about how to proceed with research. While that point could have been made using hypothetical discoveries in physics, the digression into biology has allowed us to examine the issues in the context of a live debate.

5.6   A Last Word on Conservatism

We've already seen how conservatism works against ID, but there is another way it comes into play in all this. In my view, conservatism is also the reason ID attracts so many theists and few nontheists. If one already believes in an intelligence that can play the role of designer, then ID-based science isn't a radical move. It's plausible that the working of the Designer might be inferred from observations. Design arguments tend to fit with what most theists already believe. For atheists on the other hand, making ontological room for a designer, especially a supernatural one, requires a radical change. An atheist would have to revise not only his or her scientific views but metaphysical and theological ones as well. A far more conservative move is to reject ID out of hand or, easier still, label it creationism and ignore it altogether. Even if the evidence for ID were stronger, it might be rational for an atheist to believe that *something* must be wrong with that evidence even if one cannot say what it is.

Ironically, the same goes for young earth creationism. Creationists believe that a literal, historical interpretation of Genesis 1 is the only legitimate approach to the text. Moreover, they consider the Bible to be the best source of information about such matters. Many Creationists admit that current science does not support their view, but, much like the atheist in my previous example, they believe that there must be *something* wrong with any evidence that points to an ancient cosmos. Once again, this is just conservatism at work.

All this goes to show that background beliefs, including higher-order, philosophical beliefs, play a much greater role in our thinking than we commonly assume. We each have strongly held, generally unquestioned assumptions about the world and how it should be understood. One goal of this chapter was to show how these assumptions shape the assessment of

ID. Given the subtle way such beliefs operate, we must identify and examine them instead of simply assuming that whatever principles happen to be in play today are obviously right and true.[16] This is the kind of analysis typically done in the philosophy of science, not in science itself. And that's okay. There is a division of labor among the disciplines. Philosophers and historians are better equipped to analyze the role of shaping principles than scientists themselves. Those in the natural sciences might even find such analysis useful. If you're thinking, "that's just what I would expect a philosopher to say," consider one other opinion:

> I fully agree with you about the significance and educational value of methodology as well as history and philosophy of science. So many people today—and even professional scientists—seem to me like somebody who has seen thousands of trees but has never seen a forest. A knowledge of the historic and philosophical background gives that kind of independence from prejudices of his generation from which most scientists are suffering. This independence created by philosophical insight is—in my opinion—the mark of distinction between a mere artisan or specialist and a real seeker after truth.

Hear, hear. The author?
Albert Einstein.

Notes

1　One example of this split occurs in the literature on fine-tuning. Many theistic philosophers and scientists believe that God has in fact set the fine-tuned constants to their life-permitting values. Gingerich, Polkinghorne, and others thus believe that the theistic explanation is better than any naturalistic alternative. Nonetheless, the majority of these scholars are critics of the ID movement.

2　Dembski (2004b) has argued that Behe's irreducibly complex systems are a subset of those with complex specificity.

3　See Koperski (2003) for an overview of the many arguments surrounding ID. Ratzsch (2001) is also especially helpful.

4　Pennock has more recently argued for a softer version of MN, one that does not offer a strict demarcation between science and nonscience (Pennock 2009). As I will argue, that is a wise move. Nonetheless, it is clear from the *Kitzmiller* transcript that Judge Jones understood Pennock's view as something more. Whatever Pennock had in mind, it came across as strict demarcation. See Monton (2006, 2009, 47–58) for more on demarcation issues in *Kitzmiller*. Philosopher Michael

Ruse is more willing to defend MN as demarcation (Ruse 2001, 365–372), although the boundary, he says, is littered with examples that do not clearly fit on one side or the other.

5 See Numbers (2008) for more.

6 The twentieth-century Christian fundamentalist movement was named after a series of essays known as *The Fundamentals*. Surprisingly, many of its contributors were not classical creationists. Conservative theologian Benjamin B. Warfield (1851–1921), who wrote "The Deity of Christ" in volume 1 of *The Fundamentals* said, "I am free to say, for myself, that I do not think that there is any general statement in the Bible or any part of the account of creation, either as given in Genesis 1 and 2 or elsewhere alluded to, that need be opposed to evolution" (Warfield [1888] 2000, 130). See Numbers (2007, 59–71) for more.

7 As Larry Laudan has argued, Cartesian intelligibility ("clear and distinct ideas") had to be sacrificed in order for Newtonian gravitation to be accepted (1984, 60–61). The idea that Newtonian gravitational forces act at a distance rather than by mechanical contact was highly controversial.

8 When scientists like Gauch (2003) talk about the scientific method, they generally aren't interested in the demarcation problem. They are instead looking for the most general principles shared by all the sciences: induction, deduction, probability theory, etc. While every special science no doubt uses these, there is nothing uniquely scientific about them.

9 See Ratzsch (2005), especially pages 136–139, and Sober (2011, 373–374).

10 See Numbers (2007, 39–58) for an overview.

11 Ratzsch isn't demanding direct observation here as a means of testing. Electrons are intrinsically unobservable, but there are still good reasons for believing they exist. Cosmologists believed in black holes based on papers by Hawking and Penrose in the 1960s, long before there was any empirical evidence. Even what evidence we have for black holes falls short of direct observation, the way I observe a rabbit eating my wife's flowers.

12 Also see Sober (2007, 6–7).

13 See the Discovery Institute's annotated bibliography at http://www.discovery.org/a/2640. In my view, it is difficult to see 60 papers published over almost thirty years as a thriving research program.

14 The uproar over Mayer (2004) is a case in point.

15 See, respectively, Kauffman (1995), Goodwin (1994), Shapiro (2011), Conway Morris (2003), and S.B. Carroll (2005).

16 As we saw in Section 1.3.2, appealing to the scientific method to sort this out is circular. Any such method *presupposes* a set of shaping principles. Whatever constitutes the "best method" at a given time will be derived from shaping principles. Using such a method to assess shaping principles is like using the conclusion of an argument to assess the rightness of its premises.

17 *Letter to Robert Thornton*, December 7, 1944.

References

Behe, Michael. 2001. "Molecular Machines: Experimental Support for the Design Inference." In *Intelligent Design Creationism and Its Critics: Philosophical, Theological, and Scientific Perspectives*, edited by Robert T. Pennock, 241–256. Cambridge: MIT Press.

Carroll, Sean. 2009. "Science and Religion Are Not Compatible." *Cosmic Variance.* http://blogs.discovermagazine.com/cosmicvariance/2009/06/23/science-and-religion-are-not-compatible/. Accessed June 3, 2012.

Carroll, Sean B. 2005. *Endless Forms Most Beautiful: The New Science of Evo Devo and the Making of the Animal Kingdom*. New York: W.W. Norton.

Conway Morris, Simon. 2003. *Life's Solution: Inevitable Humans in a Lonely Universe*. Cambridge: Cambridge University Press.

Cormack, Lesely. 2009. "That Medieval Christians Taught That the Earth Was Flat." In *Galileo Goes to Jail: And Other Myths About Science and Religion*, edited by Ronald L. Numbers, 28–34. Cambridge: Harvard University Press.

Darwin, Charles. 1869. *On the Origin of Species by Means of Natural Selection: Or the Preservation of Favoured Races in the Struggle for Life*. New York: D. Appleton and Company.

Davis, Edward B. 1996. "Newton's Rejection of the 'Newtonian World View': The Role of Divine Will in Newton's Natural Philosophy." In *Facets of Faith and Science: The Role of Beliefs in the Natural Science*, edited by Jitse M. van der Meer. Vol. 3, 75–96. Lanham: University Press of America.

Dembski, William. 2004a. *The Design Revolution: Answering the Toughest Questions About Intelligent Design*. Downer's Grove: InterVarsity Press.

Dembski, William. 2004b. "Irreducible Complexity Revisited." *Progress in Complexity, Information, and Design* 3 (1) (November).

Forrest, Barbara. 2001. "The Wedge at Work: How Intelligent Design Creationism Is Wedging Its Way into the Cultural and Academic Mainstream." In *Intelligent Design Creationism and Its Critics: Philosophical, Theological, and Scientific Perspectives*, edited by Robert T. Pennock, 5–53. Cambridge: MIT Press.

Forrest, Barbara, and Paul R. Gross. 2004. *Creationism's Trojan Horse: The Wedge of Intelligent Design*. Oxford/New York: Oxford University Press.

Gauch, Jr., Hugh G. 2003. *Scientific Method in Practice*. New York: Cambridge University Press.

Gazit, Ehud. 2007. *Plenty of Room for Biology at the Bottom: An Introduction to Bionanotechnology*. London: Imperial College Press.

Gingerich, Owen. 2005. "An Astronomer's View of Christianity and Science". Interview by Renee Montagne. National Public Radio. http://www.npr.org/templates/story/story.php?storyId=4490227. Accessed November 1, 2009.

Goodwin, Brian C. 1994. *How the Leopard Changed Its Spots: The Evolution of Complexity*. New York: C. Scribner's Sons.

Herschel, William. 2010. *The Scientific Papers of Sir William Herschel V1 (1912).* Edited by J.L.E. Dreyer. Whitefish: Kessinger Publishing.

Kauffman, Stuart A. 1995. *At Home in the Universe: The Search for Laws of Self-Organization and Complexity.* New York: Oxford University Press.

Kitcher, Philip. 1983. *Abusing Science: The Case Against Creationism.* Cambridge: MIT Press.

Kitzmiller v. Dover. 2005. *Kitzmiller v. Dover, 400 F. Supp. 2d 707 (M.D. Pa., 2005).*

Koperski, Jeffrey. 2003. "Intelligent Design and the End of Science." *American Catholic Philosophical Quarterly* 77 (4): 567–588.

Laudan, Larry. 1982. "Commentary: Science at the Bar-Causes for Concern." *Science, Technology, & Human Values* 7 (41): 16–19.

Laudan, Larry. 1984. *Science and Values: The Aims of Science and Their Role in Scientific Debate.* Berkeley: University of California Press.

Lycan, William G. 1988. *Judgement and Justification.* Cambridge/New York: Cambridge University Press.

Meyer, Stephen. 2004. "The Origin of Biological Information and the Higher Taxonomic Categories." *Proceedings of the Biological Society of Washington* 117: 213–239.

Miller, Kenneth R. 2009. *Only a Theory: Evolution and the Battle for America's Soul.* New York: Penguin Books.

Monton, Bradley. 2006. "Is Intelligent Design Science? Dissecting the Dover Decision." *Preprint.* http://philsci-archive.pitt.edu/id/eprint/2592. Accessed October 22, 2009.

Monton, Bradley. 2009. *Seeking God in Science: An Atheist Defends Intelligent Design.* Buffalo: Broadview Press.

Murray, Michael J. 2003. "Natural Providence (or Design Trouble)." *Faith and Philosophy* 20 (3): 307–327.

Numbers, Ronald L. 2007. *Science and Christianity in Pulpit and Pew.* New York: Oxford University Press.

Numbers, Ronald L. 2008. "Science without God: Natural Laws and Christian Beliefs." In *When Science and Christianity Meet*, edited by R.L. Numbers and D.C. Lindberg, 265–285. Chicago: University of Chicago Press.

Pennock, Robert T. 1996. "Laws of Physics." *Biology and Philosophy* 11 (4): 543–549.

Pennock, Robert T. 2001. "Naturalism, Evidence, and Creationism: The Case of Phillip Johnson." In *Intelligent Design Creationism and Its Critics: Philosophical, Theological, and Scientific Perspectives*, edited by Robert T. Pennock, 77–97. Cambridge: MIT Press.

Pennock, Robert T. 2009. "Can't Philosophers Tell the Difference Between Science and Religion?: Demarcation Revisited." *Synthese* 178 (2): 177–206.

Proctor, Robert. 1991. *Value-Free Science?: Purity and Power in Modern Knowledge.* Cambridge: Harvard University Press.

Pynes, Christopher A. 2012. "Ad Hominem Arguments and Intelligent Design: Reply to Koperski." *Zygon* 47 (2): 289–297.

Quine, W.V.O. [1951] 1980. "Two Dogmas of Empiricism." In *From a Logical Point of View*, edited by W.V.O. Quine, 20–46. Cambridge: Harvard University Press.

Quine, W.V.O. 1995. "Naturalism: Or, Living Within One's Means." *Dialectica* 49: 251–261.

Ratzsch, Del. 2001. *Nature, Design, and Science*. Albany: SUNY Press.

Ratzsch, Del. 2005. "Intelligent Design: What Does the History of Science Really Tell Us?" In *Scientific Explanation and Religious Belief: Science and Religion in Philosophical and Public Discourse*, edited by Michael Parker and Thomas Schmidt, 126–149. Tübingen: Mohr Siebeck.

Ruse, Michael. 2001. "Methodological Naturalism under Attack." In *Intelligent Design Creationism and Its Critics: Philosophical, Theological, and Scientific Perspectives*, edited by Robert T. Pennock, 363–385. Cambridge: MIT Press.

Shapiro, James A. 1997. "A Third Way." *Boston Review* 22 (1). http://new.bostonreview.net/BR22.1/shapiro.html, Accessed April 26, 2008.

Shapiro, James A. 2011. *Evolution: A View from the 21st Century*. Upper Saddle River: FT Press Science.

Singham, Mano. 2010. "The New War Between Science and Religion." *The Chronicle of Higher Education* (May 9). http://chronicle.com/article/The-New-War-Between-Science/65400. Accessed June 1, 2013.

Sklar, Lawrence. 1975. "Methodological Conservatism." *Philosophical Review* 84 (3): 374–400.

Smolin, Lee. 2006. *The Trouble with Physics*. Boston: Houghton Mifflin.

Sober, Elliott. 2007. "What Is Wrong with Intelligent Design?" *The Quarterly Review of Biology* 82 (1): 3–8.

Sober, Elliott. 2011. "Why Methodological Naturalism?" In *Biological Evolution: Facts and Theories. A Critical Appraisal 150 Years After "The Origin of Species"*, edited by G. Auletta, M. Leclerc, and R.A. Martínez, 359–378. Roma: Gregorian & Biblical Press.

Warfield, B.B. [1888] 2000. *Evolution, Scripture, and Science: Selected Writings*. Edited by Mark A. Noll and David N. Livingstone. Grand Rapids: Baker Books.

6

Reduction and Emergence

6.1   Nothing but Atoms?

I see a student walking toward campus from the parking lot. What is it that
I see? Broadly speaking, a living organism composed of various systems:
circulatory, respiratory, etc. Those systems are comprised of organs, which
are made up of cells. If we keep going, we eventually run into molecules,
then atoms, and then quarks and other particles in the Standard Model. So
then, is the student walking across campus ultimately nothing but a collection of subatomic particles?

Some say yes; he's composed of atoms, so he is really nothing but a collection of subatomic elements. But consider his red shirt. Let's say it is
made of cotton. The cotton is composed of dyed cotton fibers. If we keep
drilling down, eventually, we will run into molecules and atoms again.
Now, is the shirt red because it is composed of red atoms? No, atoms, so
chemists tell us, are colorless. So where did the color come from? The same
folks who say my student is just a bunch of atoms will say that the redness
of the shirt is due to the way that particular bunches of atoms interact with
the part of the electromagnetic spectrum we call light. Others believe that
many of the properties of people, shirts, etc., cannot be reduced to facts
about atoms and that electromagnetism is only a small part of the scientific
explanation for color.

The first group believes in *reductionism*, the view that high-level theories, laws, and complex entities can in principle be reduced to (or explained
by) lower, more fundamental levels in nature. The expectation is that

The Physics of Theism: God, Physics, and the Philosophy of Science,
First Edition. Jeffrey Koperski.

psychology will one day be reduced to neurophysiology, neurophysiology to molecular biology, and molecular biology to organic chemistry, all the way down to quantum field theory. Extreme forms of reductionism hold that eventually the reduced theories will fade away. There will be no need for psychology, they say, once neuroscience is sufficiently mature.

In this chapter, we consider the arguments for and against reductionism and then consider its main rival: emergence. As we will see, a large part of the reductionist program is generally considered a failure. Emergentists believe that higher-level phenomena have a kind of autonomy that prevents them from being reduced to more fundamental parts. Crudely put, emergentism holds that while wholes are composed of base constituents, the properties of the wholes are not fully derived from those constituents. Theists have been keen on this development since, of course, God cannot be reduced to physics. At the end of the chapter, I will present my own diagnosis of the controversy between reductionists and their foes and then briefly consider some theological ramifications.

6.2   The Rise of Reductionism

Reductionism boasts a number of success stories. No one in the ancient world would have guessed that water is H_2O, yet we have successfully reduced this common substance to its molecular parts. Other chemical reductions fill textbooks. We also now know that air temperature is just the average kinetic energy of its molecules. Molecules that are more energetic feel warmer to us. In fact, the whole of thermodynamics is often thought to have been reduced to statistical mechanics and the behavior of atoms. Sound has been reduced to compression waves moving through the air. Color and light have been reduced to electromagnetic radiation. Ray optics has been reduced to wave optics. Observable biological traits have been reduced to genetics and DNA. The list goes on. (If you have objections to any of these examples, keep reading.)

So what exactly do we mean when we say that one phenomenon "has been reduced to" something else? The answers generally fall into two categories.

6.2.1   *Ontological Reduction*

Reduction is usually thought of in ontological terms: in some sense, the world is nothing but its most fundamental constituents. Once you understand the role of atoms in the three states of matter, it is easy to think that physical stuff is nothing but collections of atoms. Macroentities along with their properties and causal powers are just aggregates of microentities and their properties/causal capacities.

A slightly weaker version of ontological reduction says that, instead of a whole being *identical* to a collection of its parts, the lower-level entities and laws *determine* the higher without being identical to it. So while a given pain might not be identical with a particular brain state, it is causally determined by that state. On this view, underlying chemical principles cause biological phenomena. Underlying physical laws likewise cause what we see at the level of chemistry. In general, what happens at higher levels is caused by the behavior of entities at some lower level.

6.2.2   Theory Reduction

Another kind of reductionism is epistemic rather than ontological. It begins with the idea that law statements, theories, and models are constructs that scientists create in order to understand nature. Each can be accurate to varying degrees, but they are all tools used to explain and make predictions. Law statements and the rest belong to our side of the gap between beliefs about natural phenomenon and nature itself.

Theory reduction, then, has to do with how theories and law statements fit together and how we organize our knowledge about nature. It is a view about how all of the tools and structures of science relate to one another. The core idea is that we can (in principle) always explain high-level macrophenomena in terms of lower levels. As science progresses, we will continue to see how more fundamental levels of reality give rise to what we experience. This is often a matter of seeing how the whole is constituted by its parts.

Theory reductionism assumes that ontological reductionism is true. We are able to explain the macro in terms of the micro *because* nature is ultimately nothing but aggregates of the micro. Over time, reductionists expect that entire theories and even branches of sciences will be reduced to others: classical mechanics to quantum mechanics, psychology to neuroscience, etc. Many scientists would say that reduction is what scientific progress is all about. While reductionism is in part a prediction about how science will mature, this optimism is grounded in successful theory reductions over the last two centuries.

Let's now consider whether that optimism is warranted.

6.3   Popping the Reductionist Bubble

Philosophers of science have been chipping away at reductionist claims for a couple of decades now. It turns out that many of the success stories mentioned in the previous section are incomplete.

6.3.1   *Temperature to Average Kinetic Energy*

The claim that temperature has been reduced to the average kinetic energy of molecules is only true for gases. The temperature of the light bulb in my office is a different matter altogether. More generally, thermodynamics has never been fully reduced to statistical mechanics. Many of the dots have stubbornly refused to be connected (Sklar 2009). Textbook examples of thermodynamic reduction assume that containers of gases have infinitely many particles, an idealization known as the "continuum limit," and most of the examples only apply to systems in equilibrium—a very special state. Nonequilibrium systems evolving under the second law of thermodynamics have directionality: from low entropy to high. How to cash out that directionality in terms of statistical mechanics is still a matter of debate. Considering these and a number of other gaps, philosopher of physics Robert Batterman argues that statistical mechanics *cannot* account for all of the observable behavior of macroscopic systems governed by thermodynamics:

> The upshot is that the statistical mechanics of finite systems is explanatorily insufficient. While it gets the ontology of blobs of gases and fluids right, [i.e.] they are composed of a finite number of interacting molecules, there remain macroscopic phenomena—universal patterns of behavior—that cannot be explained by this fundamental theory. (2010, 1033–1034)

In this case, the more fundamental theory is statistical mechanics; the phenomena supposedly reduced are thermodynamic. The point is that it is highly questionable whether thermodynamics should be the poster child for reductionism.

6.3.2   *Classical Mechanics to Quantum Mechanics*

An important mismatch between classical and quantum mechanics is that the latter does not seem to allow for chaos (Section 4.4.3).[1] Quantum mechanics imposes restrictions on the ways systems can evolve. Electrons cannot absorb or emit any amount of energy in an atom, for example, but only discrete packets or "quanta." While these restrictions provide stability to atoms that classical models lacked, they also prevent quantum mechanical models from displaying the extreme randomness found in chaos. If we try to model the orbits of Saturn's moons, one with classical mechanics and another with quantum mechanics, the two models will diverge (Zurek 1998, 3694). The chaotic tumbling of the moon Hyperion cannot be accurately tracked using quantum mechanics. In any case, there is no clear sense in which classical chaos has been reduced to quantum mechanics.

6.3.3   Chemistry to Quantum Mechanics

While chemistry incorporates parts of quantum mechanics, it has not been reduced to quantum mechanics. Not even the shape of molecules can be derived from quantum mechanics alone.

6.3.4   Phenotype to Genotype

News about medical research often leaves the impression that it's all about the genes: biological properties are determined by genetics. In other words, one's phenotype—the collection of observable properties of an organism—is driven by one's genotype, an organism's genetic makeup. Phenotype has been reduced to genotype, or at least that is the impression.

This conclusion runs counter to cell research over the last decade. It is now clear that an organism's traits are due to far more than just the expression of genes. Genes are an important part of the biological story, no doubt, but there are many more chapters to that story than researchers believed a generation ago. Evolution is not, as biologist/philosopher Massimo Pigliucci writes, driven by the "selfish gene":

> For instance, *pace* [Richard] Dawkins, it is becoming increasingly untenable to hold a "genecentric" view of the evolutionary process (especially considering that new discoveries in molecular biology keep questioning the very meaning of the term "gene"), and the classic textbook definition of evolution as a change in gene frequencies … simply does not begin to account for what evolution actually is. (Pigliucci 2007, 2746)

Pigliucci is referring in part to research in *epigenetics*: layers of inheritance that go beyond the gene. These mechanisms, which regulate the expression of genes, can change within a single generation, rather than over long

Reduction and Emergence

stretches of evolutionary time. For example, the diet and environment of parents can influence the lifespan of their yet to be conceived children. Epigenetics is just one way biological research is moving beyond the gene. In general, recent trends in biology have not been helpful to genetic reduction (Brigandt and Love 2012).

6.3.5 Ray Optics to Wave Optics

Ray optics is what one learns in high school physics: light represented by rays undergoing reflection and refraction. Wave optics is a more fundamental theory based on the fact that light is part of the electromagnetic spectrum. It's surprising, therefore, that the equations used to bridge wave and ray optics break down when describing some of the theory's most important phenomena: caustics (Batterman 2002, 88). A caustic is a broad phenomenon that includes focal points. As light reflects off a curved surface, like the edge of a coffee mug (Figure 6.1), it is concentrated along the brightly lit band. The wave equation used to model the behavior of light works well until it runs into a caustic.[2] At that point, the light intensity becomes infinite—a physically impossible result. The point is that while important connections have been established between the higher-level theory and the more fundamental one, it is not the case that ray optics has been fully reduced to wave optics. To understand the behavior of caustics, we need both.

Figure 6.1 Caustic.

Many reductionists hope that more research will eventually solve these problems. "We just need to know more about how the constituents work in large numbers." That seems promising so long as we're talking about material objects—things made up of molecules. However, many and perhaps most physical properties are not related in terms of parts to wholes. Waves, for example, transport energy, not matter. Most waves are transmitted through matter rather than being composed of it. A wave does not stand in a whole-to-part relationship with the material it moves through.

While some might hope that future research will rescue reductionism from these counterexamples, the next case is considered something of a slam dunk.

6.3.6   *Quantum Entanglement*

There is a simple elegance to vector analysis, even at the level of freshman physics. Once one identifies all of the contributing forces, the resultant force is a matter of geometry and vector addition. The composition of forces is a paradigm example of reductionist thought in physics. The contributions of the parts directly lead the behavior of the whole.

Many quantum mechanical systems do not exhibit this part–whole relation. The state of an *entangled* system cannot be decomposed into parts.[3] Unlike the vector analysis of forces, the "components" of an entangled system no longer have individual states. There is no mathematical or physical way to refer to discrete parts that can contribute to the whole. Entanglement thus imposes an irreducible quantum holism on physics, as philosopher of physics Tim Maudlin argues:

> Quantum holism ought to give some metaphysicians pause. … [One] popular "Humean" thesis holds that all global matters of fact supervene on local matters of fact…. Once the local facts have been determined, all one needs to do is distribute them throughout all of space-time to generate a complete physical universe. Quantum holism suggests that our world just doesn't work like that. The whole has physical states that are not determined by, or derivable from, the states of the parts. Indeed, in many cases, parts fail to have physical states at all. The world is not just a set of separately existing localized objects, externally related only by space and time. Something deeper, and more mysterious, knits together the fabric of the world. (Silberstein 2002, 97)

Entanglement is an especially damaging counterexample to reductionism. Quantum mechanics is a fundamental theory. There is no scenario in which this quantum holism will be reduced away by some future theory.

Notice that many of these examples are within physics itself. They are not matters of our ignorance of a messy world and are far from perennial questions about the reduction of mind to body. The overall state of science, according to physicist Michael Berry, is not what reductionists had envisioned:

> Our understanding of the world is a patchwork of vast scope; it covers the intricate chemistry of life, the sociology of animal communities, the gigantic wheeling galaxies, and the dances of elusive elementary particles. But it is a patchwork nevertheless, and the different areas do not fit well together. (Berry 2002, 41)

Science pushes on, of course, but the winds are not blowing in the direction of reductionism.

The reductionist program also fell out of favor among philosophers toward the end of the twentieth century, although this trend had very little to do with science. Donald Davidson's argument for *anomalous monism* was particularly influential (Davidson 1970). The idea is that while everything is made up of physical stuff, phenomena at different levels cannot always be linked together in lawlike ways. *Supervenience* became the buzzword of choice: higher-level laws and phenomena supervene on lower levels, but cannot be reduced to those lower levels. Note that supervenience is something weaker than causation. The upper levels still depend/supervene on the lower without being directly caused by them. Molecular biology, for example, rests on a biochemical foundation. If the former supervenes on the latter, then there is no change in biological properties without some change in the underlying chemical properties.[4] Supervenient properties depend on a subvenient base. What kind of "dependence," exactly? Well, that's a tough question. None of the proposed answers attract much consensus, and supervenience itself has begun to fall out of favor.

Before we get to an alternative, let me offer one last word on reductionism. I have been leaning hard in this chapter in the other direction, emphasizing its failures. That is not to say that analysis of component parts and underlying causal structures has not been fruitful. Genetics did arise from a generally reductionist way of thinking. So did molecular chemistry, statistical mechanics, and many other special sciences. No one wants to do away with part–whole analysis. Let's just not make it synonymous with the scientific method. Reduction is a methodological shaping principle.[5] Looking to underlying mechanisms is often, but not always, useful. But reduction is like any other shaping principle in science: researchers should ride it as far as they can, then ditch it when it no longer proves useful.

Let's now turn to a more recent approach.

6.4   Emergence

Emergence is roughly the inverse of reduction: wholes are more than the sum of their parts. The reason we have all those examples of failed reduction, say, proponents of emergence, is that *new* entities, laws, and/or properties sometimes emerge from a base level. Three things characterize emergence (Batterman 2009). First, if A emerges from base level B, then A-level phenomena cannot be reduced to B level. Second, the behavior of A-level wholes cannot be predicted from knowledge of its B-level parts. Third, there is a novelty in the emergent whole. What we see at the A level is new and unexpected given its base.

Emergentists do not take the division of labor we see in the sciences as merely pragmatic. The sciences operate at different levels, from fundamental particles through biological systems to societies and economics, because nature organizes itself that way. The emergence of new levels is part of the structure of reality, they argue, not something we impose on it to facilitate understanding.

Examples of emergence begin with the failed reductions in the previous section. Both chaotic dynamics and the shape of molecules emerge from the physics of quantum mechanics, but neither can be reduced to it. Robert Bishop cites hermaphrodite clownfish as another example (personal correspondence). In one species, young clownfish are born male. If a dominant female dies, a male will (amazingly) change its organs into female ones. This phenomenon cannot be wholly reduced to matters of biochemistry. The switch itself is induced by the male/female imbalance within the clownfish social structure. One cannot explain the cause and effect relation in terms of the biochemical makeup of the fish alone.

The first use of emergence was in reference to consciousness. Most mind–body dualists believe that a mind could be conscious apart from its body. Materialist views of the mind, on the other hand, hold that a brain and body are necessary for consciousness. For the materialist, consciousness seems to emerge in a novel and unexplainable way from its neurophysiological base. Even some dualists agree. William Hasker (1999) has argued that minds emerge from brains in an analogous way to how fields emerge from magnets. The field depends on the magnet, but is not identical to it.

Finally, as Aristotle made plain with his four forces, purposeful behavior cannot be explained by physics alone. John Polkinghorne often uses the example of a boiling teapot. One can explain the phenomenon in terms of chemistry and phase changes, but a full explanation requires more: I want to make some tea. Chemistry alone cannot explain why there was water put in the pot in the first place. The same goes for the motion of the airplane approaching our local airport. Physics describes its behavior in terms of combustion, lift, drag, etc. Physics cannot explain why those very passengers are on that plane, however. We need intentions and beliefs for a full explanation.

We should make a distinction here. As is the case with reduction, there is both an ontological and epistemic way to understand emergence. For the most part, I have been discussing emergence as an ontological matter: new properties, causes, laws, and entities come into being when the conditions are right. Lower levels provide the necessary conditions for higher-level phenomena, but the lower does not determine the higher. On an epistemic reading of emergence, the appeal to higher levels is merely pragmatic. Epistemic emergence is about law statements, theories, states, and concepts. Strictly speaking, nothing new pops into reality in cases of epistemic emergence. We appeal to higher-level laws and macro entities as a matter of practical convenience. Emergent phenomena are irreducible only in the sense that theory reduction has failed, perhaps because there are just too many molecules to keep track of.

Ontological emergence quickly leads to questions about causation. What kind of causal powers do emergent entities have? Certainly, they have the ability to influence other entities at their same level: molecule to molecule and organism to organism. But ontological emergence also seems to entail *downward* causation in which the higher levels influence the behavior of lower ones. Mental states are a prime example. Consider the causal chain starting with my intention to drink from a coffee mug and ending with the motion of my hand and the mug. The content of the mug is a matter of chemistry—mostly water molecules. Intentions are high-level emergent properties.[6] Here is a case of emergent properties causing changes at a more fundamental level.

Of course, reductionists say it's just a matter of time before all these things will be explained and the full hierarchical unity of the sciences restored. Once we know enough about the brain, they say, then mental causes will be reduced to neurophysical ones. Perhaps. Then again, they have been saying that for two centuries.[7] At what point should we stop accepting promissory notes? Even most naturalistic philosophers have now given up on full-blown theory reduction, settling for ontological reduction instead. In other words, they still believe that ultimately everything is made up of subatomic particle but have given up hope that science will be able to complete the reduction of levels.

Speaking of promissory notes, it is time to deal with one made in Chapter 4 regarding divine action (Section 4.4.4.3). A recent trend among noninterventionists is the appeal to emergence and downward causation (Peacocke 1993, 53–55; Russell 2008, 124). Although these proposals are often vague and based on analogies, the core idea is that God's influence on the cosmos is from a high—and perhaps the highest—level in this hierarchy. On this view, God acts in a top-down way analogous to the intention-to-moving-coffee-mug example. If mental states can cause physical changes in the world without violating any laws of nature, then God's intentions can do the same. Downward causation is just as available to God, on this view, as it is to us.

While the tide is slowly turning away from reductionism, the new emergence program has problems of its own.

6.5   Problems and Puzzles

Philosopher Jaegwon Kim has been a central figure in matters of reduction, supervenience, and emergence in the last 15 years. He presents three related puzzles for the antireductionist (Kim 2008).

6.5.1   Causal Redundancy

Consider mental states and brain states, again. Call the sensation of biting into a jalapeno pepper M1 for mental state one. This naturally causes a desire M2 for something to drink that will cool the burning of M1. The horizontal arrows in Figure 6.2 link causes to their effects. The vertical arrows represent emergence, not causes. Let B1 and B2 be brain properties

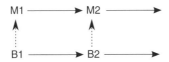

Figure 6.2   Levels diagram.

from which M1 and M2 emerge. (Alternatively, one could think of the entire matrix of mental causes emerging from a lower level rather than individual mental states arising from specific brain states.)

Psychology works at the upper level, discerning how certain mental states give rise to others, for example, how a traumatic experience in the past influences one's current choices. Neuroscience focuses on matters of the brain, for example, how hormonal changes affect the firing rate of neurons. Kim's main challenge is that on this view there are two sets of competing causes. Say that the desire M2 leads to an intention to drink something cold (M3). There is a string of mental causes and effects beginning with M1. There is another string of causes and effects, however, operating at the level of my brain/body. This one (B1, B2, B3, etc.) does not depend on mental states.

The problem is that mental causes seem to be redundant. If mental states somehow depend on brain states and yet brain states have their own network of causal relations, what work can mental states actually do? It would seem that the upper track of causes, M1 → M2, is superfluous, given the lower and more fundamental track, B1 → B2. Perhaps, we should drop the causal arrows in the upper level, thus eliminating any tension between base causes and emergent ones (Figure 6.3). Epiphenomenalism is the view that while mental states are real, they do not have any causal powers. The causal work is done at the next level down.

Epiphenomenalism resolves the tension between competing causal tracks, but at a stiff price. If mental states do not have any causal powers, then they literally cannot bring about any changes. I might intend to drink some cold water, but that intention cannot cause my hand to move. Intentions, *qua* mental states, cannot cause anything if epiphenomenalism is true. Desires and intentions merely ride along atop of the real causal network at the level of brain states. That, most agree, is not a good solution. Few are willing to accept the idea that our mental worlds are causally cut off from the rest of the reality. We are not merely passive receivers of sense data with no ability to actively respond. Rejecting epiphenomenalism, however, leaves the problem of causal redundancy unresolved.

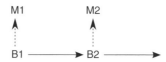

Figure 6.3   Epiphenomenalism.

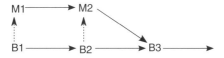

Figure 6.4   Overdetermination.

6.5.2   Downward Causation

This problem is closely related to the first. Say that I have just bitten into that jalapeno. Let M2 be my intention to reach for a glass of water. The diagonal arrow in Figure 6.4 represents the downward cause from M2 to the nerves in my brain (B3) that will get my hand moving. However, M2 is not the only cause of B3. B2 is as well. If we allow for downward causation, then there are two causes for one event, a case of causal overdetermination. One of the two causes, once again, seems to be unnecessary, and yet every case of downward causation directly leads to overdetermination. This is a bigger problem than redundancy because of the causal logjam at B3.

6.5.3   Causal Drainage

The first two problems imply that either there are no mental causes or, if there are, they can't do anything. All of the real causal action takes place at the next level down. Mental causes drain away into the level below. But why stop there? There are many levels below that of neurophysiology. Since the network of neural events itself emerges from a biochemical base, causal redundancy and overdetermination apply to brain states as well as mental states. There is nothing to keep neurophysiological causes from draining away into more fundamental biochemical ones. And since biochemistry emerges from atomic physics, the sequence repeats itself. At the end of the day, it would appear that the only real causal work in nature is done at the most fundamental level of particle physics—if there is one.

This is a radical conclusion. It means that virtually every causal claim ever made is false. There are no real causes in chemistry, biology, astronomy, or baseball. Macroscopic causes have all been drained away, entailing widespread causal antirealism in every area of life except physics. The only level of nature left with causal power is the one at the bottom, as Paul Humphreys argues:

[Only] the most basic physical properties can be causally efficacious if these arguments are correct. Indeed, unless we have already isolated at least some … fundamental physical properties, every single one of our causal claims within contemporary physics is false and consequently there are at present no true physical explanations that are grounded in causes. (1997, 3–4)

Well, that's not good. Isn't there some way to escape both reductionism and blanket antirealism about causes?

Fortunately, science itself can point the way between these two extremes.

6.6   Physics, Causes, and Levels

Let's try to find a solution to some of these problems. A conservative approach starts with the notion of causation. 'Cause' is a notoriously ambiguous term, and most philosophers believe that these problems stem from these ambiguities. They think that the approach illustrated by the levels diagram (Figure 6.2) is basically right. All we need are more restrictions on how the causal arrows work. The problems can be fixed, in other words; we just need to think harder about causes and levels. Patch up the holes in the boat rather than abandon ship.

Others are beginning to adopt a more radical view. They believe that science itself—especially physics—has been forced to deal with these same issues but has taken a different approach. Perhaps, what we need is to stop thinking that nature coalesces into a single hierarchy of discrete levels.[8] The division of labor in the sciences is fine: physics and chemistry tend to focus on the small; biology and social sciences on the large. Causes, on the other hand, are not restricted to fixed levels as if they ran along a stacked freeway. Recent work, especially in nonlinear dynamics, favors "nested and entangled" systems over levels thinking. Philosopher Michael Silberstein bluntly argues that we should reject

> the layered model of reality as divided into a discrete hierarchy of levels. The universe is not ordered as a hierarchy of closed autonomous levels such as atoms, molecules, cells, and the like. Rather, the universe is intrinsically nested and entangled. The so-called physical, chemical, biological, mental, and social domains of existence are in fact mutually embedded and inextricably interconnected. That is, mental properties are not on a higher level than neurochemical properties, the former are not on a higher level than chemical properties, and so on. It is best to view the word as divided into systems and subsystem, not levels.... (2006, 204)

On this view, there is strictly speaking no downward causation, at least not in terms of levels. The actual situation is messier: constraints, order parameters, degrees of freedom, feedback, and even teleology.

Consider a simple example. Take a large pan of water on a hot stove. Given the difference between the temperature of the burner and that of the open air at the top, the water at the bottom of the pan expands and becomes less dense. If the heat is sufficient, the fluid will rise and begin to rotate within stable (Rayleigh–Bénard) convention cells, from bottom to top and then down again. As Bishop points out (2005, 233–238), the fluid is obviously needed in order for there to be convection—no fluid, no convection cell. *Contra* the reductionist, the local interaction of the water molecules themselves is not enough to cause the water to self-organize in this way. To explain convection, one must include properties that cannot be reduced to the properties at the level of molecules. A host of nonlocal factors is needed: "temperature gradients, gravity, long-range forces and correlations, physical boundaries and symmetries, conservation laws among other things are also involved" (Bishop 2005, 238). In other words, there are causally relevant features of convection apart from the local contact forces between molecules. It is not the case that the behavior of the whole (convection cells) can be explained in terms of the causes acting at the level of the parts (molecules) alone.[9]

We are exploring the idea that (i) an overly metaphysical understanding of levels in nature is responsible for the problems in the previous section and (ii) the way physics deals with these issues is somewhat more subtle. Let's discuss the second part a bit more. The headline-grabbing discoveries are usually made at very large scales, like astrophysics, or very small ones, like the Standard Model. The less flashy areas of research are somewhere in the middle. Fluid mechanics deals with substances at a familiar scale. We are comfortable with the idea that air flows over a wing or that water can run in both simple and turbulent ways. Surprisingly, fluid mechanics does not treat fluids as being composed of atoms. Matter is treated as a continuum without gaps. The molecular nature of matter is ignored.

Given that molecules exist and matter is not a continuum, how can physicists and engineers just ignore the facts? One of the lessons of macroscale physics is that the nature of phenomena at smaller scales is often irrelevant. The details about, in this case, molecular interactions can safely be ignored. In fact, all sorts of different microconfigurations give rise to the very same macrophenomenon. The precise nature of the constituents makes no difference at larger scales, as physicist George Ellis explains:

> In general many lower level states correspond to a single higher level state, because
> a higher level description … is arrived at by ignoring the micro-differences

> between many lower level states … and so throwing away a vast amount of lower level information (coarse graining). (2006, 86)

As far as aeronautical engineers are concerned, air could be made up of atoms, Newtonian corpuscles, Boscovichian point masses, or Leibnizian monads. So long as air behaves a certain way at the scale of an airplane wing, what the molecules are doing is irrelevant. The very same equations emerge regardless of whether matter is molecular or continuous (Truesdell 1984, 55).

So in one way, the physics of fluid mechanics is simpler than the levels approach would indicate. The behavior of the whole does not depend on the precise state of its parts. Microdetails are safely ignored when it comes to the behavior of the whole. There is another sense, however, where the behavior of midscale bodies demands far more sophistication than the levels diagram can support. Let's consider an example.

Continuum mechanics is the equivalent of fluid mechanics for solids. It is the physics of bending beams and elastic balls. The fundamental law of continuum mechanics, known as Cauchy's first law of motion, applies to all midscale bodies regardless of material composition. In this sense, it is a completely general law. One needs to supply Cauchy's law with initial conditions in order to solve it—no surprise there. But it also requires something less familiar, namely, *constitutive relations*. These mathematical relations differ from material to material and capture the way in which causal signals move through a body. In solids, constitutive relations specify how a body will respond to pressure and tension. While this might sound simple enough, it requires some very sophisticated mathematics. (The details of tensor analysis are beyond what engineers are exposed to as undergraduates.)

Constitutive relations capture the necessary material conditions for observable phenomena. There can be no compression wave passing through a rod without the iron that makes up the rod itself. But that's only the beginning. An array of causally relevant information is needed, all from sources other than the material base. Even for something as simple as a plucked string, one needs initial conditions (where and how much of a "pluck"), boundary conditions (how the string is connected to the body), body forces (such as gravity; it affects the entire string), and constitutive relations in order to solve Cauchy's law.

The point is that the relation between part and whole in midscale physics does not look like the levels diagram. Causally relevant information comes from the environment and other macroscale sources, all of which is mathematically encoded in different ways. There is no one-size-fits-all causal arrow that moves a system from state to state.

Reductionists have mixed feelings about all this. They are happy to see the emergence program struggle with causal redundancy and downward causation. They would prefer that we give up emergence altogether and return to the fold. And they are not impressed with these appeals to midscale physics for guidance. In their view, continuum mechanics can in principle be reduced to atomic physics and atomic physics to quantum mechanics. All this talk of emergence and causally relevant information is merely pragmatic. We have imperfect knowledge about how to derive macrophysics from microconstituents, they argue, but research continues. For now, reductionists would have us treat emergence an artifact of our ignorance rather than a metaphysical truth. In other words, there is a gap between epistemic and ontological emergence. Reductionists might grudgingly allow loose talk about the former, but they completely deny the ontological version. These examples simply reflect a limited understanding of nature, they say. So while reductionists believe that we need continuum mechanics for engineering purposes, they claim that is merely for mathematical convenience. The so-called "emergent phenomena" have no metaphysical standing.

Very well, but one might equally wonder how reductionism itself was able to bridge this gap between our limited epistemic access to nature and nature itself. After all, reductionists have always pointed to success in science as evidence of ontological reduction. "Temperature is nothing but mean kinetic energy, light is nothing but photons, etc." Yet thermodynamics is merely a theory—something we construct. The same for the molecular theory of matter and electromagnetism. The reductionist's argument about the gap between epistemology and metaphysics applies equally well here. Why is reductionism not seen as merely pragmatic/epistemic, merely a relation between theories? The ontological reductionist is naively realist about theories and intertheoretic relations when reductionism seems to work but turns to skepticism and antirealism when faced with examples of emergence.

As we saw earlier, reduction in science has worked and the reductionist plays up these success stories. He wants reduction to be a metaphysical shaping principle, not just a methodological one: good science tries to reduce because that's the way reality is. But insofar as reductionism gets its support from science, that support is deteriorating. Hence the rise of emergence. Of course, the reductionist does not want us to be realists about those examples. He can only be fully realistic about high-energy physics—the most fundamental level. Everything else must ultimately reduce to the ground floor of reality.

At the very least, there should be some parity here between reduction and emergence. Both draw on science to support their case. Reductionism ought not to be the default position. In my view, both have earned a place in science and the philosophy of science.

6.7   Theology and Emergence

The limits on reduction in physics that we've seen here should tip the balance away from naive optimism about reduction elsewhere. There is now less reason to think that mental states and free will can be fully reduced to neuroscience or that religious truths can be reduced to sociology, psychology, and evolutionary biology. Emergence thus has a natural appeal to theists as an intrinsically antireductionist program. Insofar as religious beliefs are true, they cannot be reduced to scientific facts.

More narrowly, emergence is often embraced by theistic evolutionists. If emergence is ubiquitous in nature, then the development of new creatures through evolution looks like part of a general principle within creation. Physicist Howard Van Till (1998) argues that nature was given a "robust formational economy" at creation. All of the creatures, structures, and systems that God intends were in a sense embedded in the Big Bang itself. This idea is not entirely new. According to Augustine,

> It is ... causally *(causaliter)* that Scripture has said that earth brought forth the crops and the trees, in the sense that it received the power of bringing them forth. In the earth from the beginning, in what I might call the roots of time, God created what was to be in times to come. (1982, 153)

This is not to say that Augustine believed in either evolution or emergence. The point is that the broader idea of God creating nature in such a way that it would have the capacity to "bring forth" (Gen 1:11–12) new entities is an ancient one.

Noninterventionists generally believe that emergence fits well with their perspective on divine action. Arthur Peacocke endorses a view in which God acts on the whole of creation, thereby influencing its parts, analogous to heating a pan of water to induce the water molecules to organize into convection cells (see Section 6.6). God need not intervene within the lower levels of reality; special divine action is holistic and top-down (2006, 261–265). Peacocke's view is, I believe, more of a gesture in a certain direction than a concrete proposal, so it is not clear how much support it derives from research on emergence. (What exactly is the "whole" or "top level" at which God is acting and how is it connected to the lower levels?) Peacocke seems to presuppose a tight hierarchy of levels and ubiquitous downward causation. Otherwise, there is no way for God to influence events at the level of, say, biochemistry from the highest level—whatever that happens to be. As we have already seen, however, the hierarchy-of-levels metaphysic generates its own set of problems.

Overall, the failure of reductionism is more important for theism than is the success of emergence. If every entity, cause, and law could be reduced to the next lower level, all the way down to fundamental physics, then there would not be much room left for theism. Religion itself would be reduced to psychology, sociology, and evolution—a reduction oft claimed by naturalists. The failure of wholesale reductionism in the physical sciences casts some doubt on the easy reduction of religion to socio-psychological phenomena. At the very least, it would be refreshing if the social sciences occasionally took religious belief to be a mixture of practice *and* knowledge claims rather than a quaint psychological appendage left over from our evolutionary past.

Notes

1 This is yet another problem for those trying to use chaos to solve the amplification problem back in Chapter 4 (Section 4.5.5). See Koperski (2000, 553–556) for more.

2 Technically, this statement is about the interfering wave sum, an equation used to model the behavior of light at very short wavelengths.

3 See Humphreys (1997) for more on quantum entanglement as a counterexample to reductionism.

4 This is just one variety of supervenience. See McLaughlin (2011) for more.

5 One could add methodological reductionism as a separate category to ontological and theory reduction as a way of capturing this idea, as Rodney Holder has suggested (private correspondence).

6 I do not mean to beg the question against mind–body dualism here. As Hasker shows (1999), one can be a dualist and believe in the emergence of mind. Many materialists, as Thomas Tracy points out (private correspondence), are still hoping that mind can be reduced to brain and that emergence talk will prove unnecessary.

7 As Baron d'Holbach put it, "*Will*, is a modification of the brain…" (1835, 58). A modern reductionist would insert the word 'just' after 'is.' Other than that, not much has changed.

8 Some, like Thalos (2013), want to do away with the notion of levels altogether. Others, like Love (2011), take a less radical approach, arguing that the levels recognized in different areas of science cannot be put together into a single hierarchy.

9 As Bishop notes, examples like these also undercut the causal closure of physics discussed in Chapter 4 (Section 4.7). Without chemical and environmental causes interacting in complex ways, there would be no convection. The "nested and entangled" causes that Silberstein mentions are generally not matters of subatomic physics.

References

Augustine. 1982. *The Literal Meaning of Genesis*. Translated by John Hammond Taylor. New York: Newman Press.

Batterman, Robert W. 2002. *The Devil in the Details: Asymptotic Reasoning in Explanation, Reduction, and Emergence*. Oxford/New York: Oxford University Press.

Batterman, Robert W. 2009. "Emergence in Physics." In *Routledge Encyclopedia of Philosophy*, edited by Edward Craig. London: Routledge. http://0-www.rep. routledge.com.library.svsu.edu/article/Q134.

Batterman, Robert W. 2010. "Emergence, Singularities, and Symmetry Breaking." *Foundations of Physics* 41 (6): 1031–1050.

Berry, Michael. 2002. "Chaos and the Semiclassical Limit of Quantum Mechanics (Is the Moon There When Somebody Looks?)." In *Quantum Mechanics: Scientific Perspectives on Divine Action*, edited by Robert J. Russell, Philip Clayton, Kirk Wegter-McNelly, and John Polkinghorne, 41–54. Berkeley: Center for Theology and the Natural Sciences.

Bishop, Robert C. 2005. "Downward Causation in Fluid Convection." *Synthese* 160 (2): 229–248.

Brigandt, Ingo, and Alan Love. 2012. "Reductionism in Biology." In *Stanford Encyclopedia of Philosophy*, edited by Edward N. Zalta. Stanford: Metaphysics Research Lab, Center for the Study of Language and Information, Stanford University. http:// plato.stanford.edu/entries/reduction-biology/. Accessed August 2, 2013.

Davidson, Donald. 1970. "Mental Events." In *Experience and Theory*, edited by Lawrence Foster and J.W. Swanson, 79–101. Boston: University of Massachusetts Press.

D'Holbach, Paul Henri. 1835. *The System of Nature, Or, Laws of the Moral and Physical World*. Translated by H.D. Robinson. New York: G.W. & A.J. Matsell.

Ellis, George F.R. 2006. "On the Nature of Emergent Reality." In *The Re-Emergence of Emergence*, edited by Philip Clayton and Paul Sheldon Davies, 79–107. Oxford: Oxford University Press.

Hasker, William. 1999. *The Emergent Self*. Ithaca: Cornell University Press.

Humphreys, Paul. 1997. "How Properties Emerge." *Philosophy of Science* 64 (1): 1–17.

Kim, Jaegwon. 2008. *Physicalism, or Something Near Enough*. Princeton: Princeton University Press.

Koperski, Jeffrey. 2000. "God, Chaos, and the Quantum Dice." *Zygon* 35 (3): 545–559.

Love, Alan C. 2011. "Hierarchy, Causation and Explanation: Ubiquity, Locality and Pluralism." *Interface Focus* 2 (1): 115–125.

McLaughlin, Brian. 2011. "Supervenience." In *Stanford Encyclopedia of Philosophy*, edited by Edward N. Zalta. Stanford: Metaphysics Research Lab, Center for the Study of Language and Information, Stanford University. http://plato.stanford.edu/entries/supervenience/. Accessed August 2, 2013.

Peacocke, Arthur. 1993. *Theology for a Scientific Age : Being and Becoming—Natural, Divine and Human*. 2nd ed. London: SCM Press.

Peacocke, Arthur. 2006. "Emergence, Mind, and Divine Action." In *The Re-Emergence of Emergence*, edited by Philip Clayton and Paul Sheldon Davies, 257–278. Oxford: Oxford University Press.

Pigliucci, Massimo. 2007. "Do We Need an Extended Evolutionary Synthesis?" *Evolution* 61 (12): 2743–2749.

Russell, Robert J. 2008. *Cosmology: From Alpha to Omega*. Minneapolis: Fortress Press.

Silberstein, Michael. 2002. "Reduction, Emergence, and Explanation." In *Guide to the Philosophy of Science*, edited by Peter K. Machamer and Michael Silberstein, 80–107. Malden: Blackwell.

Silberstein, Michael. 2006. "In Defence of Ontological Emergence and Mental Causation." In *The Re-Emergence of Emergence*, edited by Philip Clayton and Paul Sheldon Davies, 203–226. Oxford: Oxford University Press.

Sklar, Lawrence. 2009. "Philosophy of Statistical Mechanics." In *Stanford Encyclopedia of Philosophy*, edited by Edward N. Zalta. Stanford: Metaphysics Research Lab, Center for the Study of Language and Information, Stanford University. http://plato.stanford.edu/entries/statphys-statmech/. Accessed September 1, 2013.

Thalos, Mariam. 2013. *Without Hierarchies: The Scale Freedom of the Universe*. New York: Oxford University Press.

Truesdell, Clifford. 1984. *An Idiot's Fugitive Essays on Science: Methods, Criticism, Training, Circumstances*. New York: Springer-Verlag.

Van Till, Howard. 1998. "The Creation: Intelligently Designed or Optimally Equipped?" *Theology Today* 55 (3): 344–364.

Zurek, Wojciech H. 1998. "Decoherence, Chaos, Quantum-Classical Correspondence, and the Algorithmic Arrow of Time." *Physica Scripta* T76 (1): 186.

7

The Philosophy of Science Tool Chest

7.1   Tools

My father was a mechanic. His son, a man of few tools, is decidedly not. I did have the good fortune, however, to live next to Bill, a retired master electrician for General Motors. When I needed a tool, I would go see Bill. When I didn't know what tool I needed, I would go see Bill. When I was hoping to fix something before my wife found out about it, I would go see Bill. Bill would stroll over to his garage, look in a couple of drawers, grab a few tools, and come over to make things right in the world once again.

Academics don't usually need tools. What we need are helpful ways to think about things—conceptual hooks to hang information on. Every discipline has ways to organize the vast of amounts of data the world confronts us with. Part of one's training in graduate school is the internalizing of the tools of one's discipline. But as Einstein knew, sometimes, helpful ways of thinking come from other areas.[1] This last chapter looks at approaches developed by philosophers of science that may be useful to those working in religion, theology, and the philosophy of religion.

Philosophers of science have spent a lot of time thinking about how theories change, what to do with surprising data and conflicting explanations, and what to say when we need more categories than true and false. Sometimes, all this is hidden behind terms such as *antirealism*, *paradigm*, *verisimilitude*, and *inference to the best explanation*. In each of the following sections, we will unpack a bit of jargon from the philosophy of science and consider some analogies and applications to matters of religion.

The Physics of Theism: God, Physics, and the Philosophy of Science,
First Edition. Jeffrey Koperski.
© 2015 John Wiley & Sons, Ltd. Published 2015 by John Wiley & Sons, Ltd.

7.2   Realism and Truth

Most philosophers of science accept some form of *scientific realism*. Realists take mature scientific theories to be true or at least approximately true, where truth is understood as something like correspondence. On the correspondence theory of truth, sentences like 'Marie is on the couch' are made true by the fact that a woman named 'Marie' is sitting on a couch. The sentence would be approximately true if Marie were, say, reclining on the couch. Whether anyone knows that Marie is on the couch is irrelevant to the truth of the sentence. 'There is a dime stuck in the cushions of the couch' is either true or false regardless of what anyone believes. When chemists say that the information contained in the periodic table is true, they mean it in a correspondence sense: these elements exist and they have the properties mentioned.

Scientific realism also affirms that things like histamine reactions, ionic bonds, and quarks are part of reality even though we cannot directly detect them. Realists believe these things exist because of the role they play in our best theories. Unobservable entities are not second-class citizens in the scientist's ontology. They exist in precisely the same way coffee mugs and my wife's flower bulbs do. Quarks, neutrinos, and many particles that only physicists talk about are merely among the more exotic pieces of the furniture of reality. The fact that we can use theories involving unobservable entities to make predictions and develop new technologies, realists say, is because those theories are true or close to it.

Finally, scientific realists take one of the goals of science to be an accurate and precise conception of the physical world. As science makes progress, our explanations for physical processes are improved and refined in the sense that they become closer to the truth.

If all of this seems perfectly obvious, then you've probably never taken a course in the philosophy of science. As we will see, there are several reasons to doubt this picture. Consider the notion of approximate truth again. Let M stand for the sentence 'Marie is sitting on the couch':

$$M = \text{'Marie is sitting on the couch'}$$

This seems intuitive enough. If Marie is sitting squarely on the couch with her feet on the floor, then M is true. If Marie has just been forced off the couch by Maggie, the yappy dog, then M is false. If Marie is kneeling on the couch prodding Maggie back onto the floor, then M is approximately true. What's the problem?

One issue is the logic of approximate truth. There are formal rules of logic for sentences that are strictly true or false. If M is true and some other sentence N is true, then the compound sentence M-*and*-N is true. If one of them is true and the other false, then M-*and*-N is false. Given a complete set of such rules, one can figure out the truth value of any compound sentence if one knows whether the base sentences are true or false. These are the same rules used in the logic gates found in every piece of electronics.

Now then, what is the logic of *approximate* truth? Let

$$N = \text{'Maggie is growling at Marie'}$$

What should we say if Marie is actually kneeling on the couch but Maggie is silently considering whether to bite Marie? Both M and N seem to be approximately true, but it's less clear what to say about M-*and*-N and less clear still about *If*-M-*then*-N. The rules of classical logic no longer apply. Without such rules, however, we can't say which inferences are valid and which are not. Critics argue that approximate truth is therefore a misnomer. Sentences, they argue, must be either true or false. There can be no third category of approximate truth.

While some claim that advances in computer science, especially involving the so-called "fuzzy logic," have answered these worries (Hajek 2010), others argue that this is only one of many problems for approximate truth (Stanford 2003). For my part, I see no way to avoid using something like it, whatever the name. Special relativity is not true *simpliciter*. It is an idealized special case of general relativity that does not strictly apply to a world with gravity. However, it would seem odd to say that special relativity is false. After all, Einstein's famous equation for mass–energy equivalence, $E = mc^2$, was derived from special relativity. To call special relativity false lumps it in with alchemy and the humoral theory of medicine.[2] Somehow or other, a more fine-grained evaluation is needed. To this end, Karl Popper preferred *verisimilitude* to distinguish it from simple truth. Ronald Giere (1999) uses the concept of *fit* but restricts it to models rather than theories or laws. Borrowing a term from recording technology, we might think of special relativity, classical electromagnetism, etc., as having some degree of *fidelity* to nature. High fidelity in music indicates that a recording more accurately captures a sound than one of lesser fidelity. Fidelity comes in degrees. The red plastic record player I owned in preschool had very poor fidelity. Cassette recording was much better. Music captured on a compact disc is better still. In this sense, special relativity has less fidelity than general relativity, but neither is simply true or false.[3]

Critics will say that fidelity is a metaphor and demand that it be cashed out in more precise terms. That's fair. For our purposes, though, I will simply maintain that scientists need some way to describe the relative "rightness" of their theories. In that regard, they are no different from philosophers, physicians, and bakers. We all have sets of beliefs with relative degrees of fidelity about the world around us. The classical categories of true and false are often too heavy-handed to capture this subtlety.

In any case, this worry about the nature of truth is only the first in a long line of problems for scientific realism. Let's consider a few more and then see what some of the other options are if realism finally seems untenable.

7.3   Antirealism

For many scientists, questioning realism is like questioning algebra: How can it not be right? While I understand their incredulity, the antirealist challenge cannot be dismissed with the wave of a hand. There are several reasons to believe that science does not provide an accurate and precise understanding of the nature of physical reality.

7.3.1   Internal Conflicts

Both quantum mechanics and the general theory of relativity (GTR) count as mature theories. They are the bases for superconductivity and lasers on one hand and the discovery of the Big Bang and black holes on the other. According to the standards of scientific realism, we should take both as true. Unfortunately, as we saw in Chapter 3 (Section 3.3.2), both cannot be true. STR and GTR deny that there is a fact of the matter whether two events are simultaneous; quantum mechanics demands it. GTR treats gravity as a purely geometrical property of spacetime; quantum field theory treats it as an exchange of particles. Such tensions between fundamental theories show that something must be wrong with one or both of them. Two claims cannot both be true if they contradict one another. Hence, one of the two theories (at least) is providing a useful set of equations, but not a literally true picture of reality. Which one? Well, that's the big question. What the antirealist wants us to recognize is that both theories cannot be true in a correspondence sense. Moreover, physics gets along just fine even though one or more of its best theories are not literally true.

7.3.2   External Tensions

The conflict between GTR and quantum mechanics is "internal" in the sense that both are theories within physics. Tensions can also arise across disciplines. As we saw in Chapter 3, standard spacetime models seem to preclude an objective passage of time. Biological systems, on the other hand, are intimately connected with time, as physicist George Ellis argues (Section 3.4.2). We need not rehearse that debate here. The point is that different sciences appear to treat time in incompatible ways. Chaos theory presents another such tension. While chaotic dynamics have been observed in astrophysics, chemistry, biology, and economics, it isn't clear how chaos is possible in a world governed by quantum mechanics (Section 6.3.2).

Whether these tensions can be resolved or not is irrelevant when it comes to the antirealism debate. When two sentences contradict each other, then at least one of them is false. The same goes for textbook theories. When two mature theories conflict, at least one of them must be wrong; we cannot interpret both realistically. But as the antirealist happily points out, this does not mean that scientists stop using those theories. Theories need not be true descriptors of reality in order for them to be useful.

7.3.3   The Success of False Theories

The most troubling antirealist argument is grounded in the success of theories that we now know are false. Under the caloric theory developed by the French chemist Antoine Lavoisier (1743–1794), heat was considered a fluid that flows from warmer systems to colder ones. Warm bodies, it was thought, have a dense supply of caloric. This theory was further developed by the French engineer Sadi Carnot (1796–1832), the father of thermodynamics. Carnot used caloric theory to correctly describe the relation between heat and mechanical power. This eventually led to the development of the Carnot cycle, which is still the starting place for every engineer's study of the heat engine in thermodynamics. The problem? There is no such thing as caloric fluid. Heat exchange is now explained by the mechanical interaction of atoms.

This is not an isolated case. Lamarckian evolution could explain the development of new traits over time but was eventually displaced by Darwin. Nineteenth-century energeticists denied the existence of atoms but were very successful in explaining energy conversion. Phlogiston theory successfully explained many phenomena involving gases and combustion. There was even a recognized process for creating "dephlogisticated air." The reason phlogiston isn't mentioned in chemistry books is because there is no such thing. Phlogiston theory was replaced by a rival: oxygen theory. This is only the beginning of a long list of successful theories that were falsified in time.

Examples like these lead to the antirealist's "pessimistic induction." The history of science is littered with successful theories that were later falsified. Why think that we are in any better position today? It seems likely that our best theories will one day end up on the scrap heap and that physicists will eventually look back on our time the way we look back on the era of classical mechanics. If so, then realists seem to have far too much faith in current science. We should be more pessimistic, says the antirealist, about the amount of truth captured in today's journal publications. But if current science cannot be trusted as a guide to physical reality, then scientific realism becomes untenable.

Phlogiston, caloric, and the rest were thriving theories in their day, although they each had rivals at the time that explained the same set of phenomena. When two theories explain the same set of data, philosophers of science say that the theories are *underdetermined*. Such theories are empirically equivalent; there are no observations that tip the scales in favor of one rather than the other. Of course, many examples of underdetermination in the history of science were eventually resolved when new evidence came to light. But even in those cases, antirealists argue, the problem has merely been obscured. Just because scientists have a clear favorite at the moment doesn't mean that there are not equally good explanations that they have not yet thought of. That there is no viable, rival theory to textbook theory *X*, in other words, might merely be due to the scientific community's lack of creativity. A better explanation might be bubbling up in the mind of some graduate student as we speak. Realists therefore live under the constant threat of "unconceived alternatives," as philosopher P. Kyle Stanford calls them (2006, 19). At any moment, what we consider the best theory of *X* might give way to another.

At this point, the reader might be wondering what other interpretations are available. If scientific realism takes a naively optimistic view of current science, where else might one turn?

7.3.4 Antirealist Alternatives

There are many different versions of scientific antirealism. According to *instrumentalism*, the goal of science is not to describe and explain the nature of physical reality. All we need from science is the ability to make successful predictions and technological advances. Whether a given law or theory is true in a correspondence sense is irrelevant; what we want is for it to work. Is there really such a thing as electric current composed of electrons? "Who cares?" asks the instrumentalist, "so long as electromagnetic theory allows us to invent new electronic devices, build power plants, and keep the lights going. Maxwell's laws work. Whether they somehow latch onto a deep truth about reality is irrelevant." Science on this view is merely a tool. Tools are neither true nor false. From this perspective, realists are trying to make science do metaphysical work that it is not intended to do.

Constructive empiricism is a more recent view proposed by philosopher of science Bas van Fraassen. He argues, like any good empiricist, that one should make a distinction between what can and cannot be observed. Van Fraassen is more or less a realist with regard to things we can see: tables, pendulums, penguins, etc. He fully believes in such objects as well as more scientific examples such as meteorites and digestive systems. Things are completely different, however, when it comes to unobservable, theoretical entities like electrons, wavefunctions, and black holes. Constructive empiricists believe that those sorts of things might exist, but there is no way to know. A good empiricist should therefore be agnostic about the existence of entities that one cannot directly detect. So while van Fraassen is fine with scientists using theories that mention purely theoretical objects, we ought to not believe that those objects literally exist.

The most influential philosopher of science of the twentieth century was a former physicist: Thomas Kuhn.[4] *The Structure of Scientific Revolutions*, published in 1962, challenged every aspect of realism. On Kuhn's view, scientists do not have direct access to physical reality. One's understanding of reality is instead mediated by whatever paradigm one is currently working under. 'Paradigm' is a technical and highly ambiguous term in *Structure*. It often means something like "scientific worldview": the collection of theories and metatheoretic shaping principles shared by the community of scientists. Every observation and bit of data a scientist is exposed to is interpreted through the reigning paradigm of the time. When scientific revolutions occur and the paradigm shifts, says Kuhn, one's overall way of thinking about reality changes as well. Scientists working under the paradigm of classical mechanics 200 years ago literally saw a different world than those working today.

To understand the challenge here to realism, let's first consider a much older view: Immanuel Kant's view of perception. For Kant, the internal world of one's experience is mediated by the senses and a set of "categories" that bring order to sense experience. The categories organize one's internal sense data in various ways including giving it three dimensions, a flow of time, and causal relations. Strictly speaking, reality itself is not three dimensional and has no flow of time or causes. The categories actively process sense data so that the world appears to be three dimensional and have a flow of time. The categories are "in us," said Kant, not "out there." Our species is in a sense hardwired with the categories in place. We have no choice but to conceptualize reality in these terms. One upshot of this is that there is no way to compare one's inner phenomenal realm with bare reality, what Kant called the "noumenal" world. One only has direct access to one's own internal phenomena. How reality is in itself, unmediated by the senses and categories, is completely unknowable.

Kuhn's view is analogous, but now, Kant's categories have been replaced by a paradigm. Having a paradigm composed of theories and shaping principles in place is a necessary condition for understanding nature. Although a paradigm is developed by the community of scientists, rather than being hardwired into each observer, it provides each person with a conceptual grid through which reality is seen and interpreted. There is no sense in which scientists can merely observe nature as it is, free from the influence of a paradigm. Moreover, there is no paradigm-free place to stand, Kuhn argued, and hence no way to compare scientific claims with reality itself. While paradigms occasionally change, no scientist is able to escape the constraints and biases of working within one set of theories or other. We are all in some sense trapped behind the veil of our conceptual schemes. If so, then the realist idea that mature theories provide an accurate picture of reality is no more than wishful thinking. What Kant did to the study of metaphysics, Kuhn does to the philosophy of science.

Social constructivism takes Kuhn one step further.[5] Philosophers have long distinguished between truths about mind-independent reality and truths that are matters of convention. Realists believe that science is about the former. The Earth revolved around the sun long before anyone realized it, and it would continue to do so if all life ceased. Matters of convention include things like traffic laws. Whether it is correct to drive on the left-hand side of the road or not wholly depends on the decisions of lawmakers. If there were no lawmakers—if we lived, say, in a Hobbesian state of nature— there would be no traffic laws. The law is a social construct.

Social constructivists reject this distinction between reality and convention. They take all areas of knowledge as social constructs: literature, philosophy, history, social science, and natural sciences. There is no discourse that is true or false in a correspondence sense. The very concept of truth is itself a construct, they say. For a statement to be true means no more than that one has successfully justified it in the eyes of one's community, whether tribal witch doctors or particle physicists (Rorty 1987). The realist notion of a mind-independent reality is itself—as you might have guessed—a social construct. Strictly speaking, there is no such thing as reality.

I have not argued for any of these antirealist positions. The goal here was merely to lay out some alternatives to realism, from the less to the more extreme.

7.3.5   Back to Realism

So the question remains, in light of all the problems for realism and the range of antirealist alternatives, why are most philosophers of science still realists of some sort or other? The main reason is the "no-miracles" argument: it would be a miracle for science to be as successful as it has been if its theories were not at least approximately true. Electrons are too small to directly observe even in principle, and yet realists believe they exist. Why? Because chemistry and electromagnetic theory have made dramatic advances based in part on the idea that there are such entities. Realists find it impossible to believe that chemistry and electrical engineering could work so well if the foundations of those disciplines were wrong. The same goes for all theories that have been thoroughly tested and produced new discoveries and that lead to engineering advances. While it's possible that such theories are wrong—we could live in a *Matrix* world, after all[6]—it is difficult to see how those theories could be false and yet so successful. Granted, there are different notions of "success" floating around; nonetheless, the intuitive weight of the no miracles argument has been difficult for antirealists to overcome.

With that said, one cannot completely dismiss the arguments against realism quite so easily. Antirealists have shown that not everything science says is true. Only the so-called "naive realists" believe otherwise. Most in the science-and-religion literature instead favor "critical realism." What exactly that means is hard to say, however, other than an acknowledgment that naive realism is, well, naive.

In order to derive a more nuanced version of scientific realism, we need to keep a few things in mind. One is that nature is messy in ways that science courses tend to ignore. There is no reason for undergraduate physics students to worry about the conflict between quantum mechanics and general relativity, the failure to fully understand turbulence, or the odd mathematical maneuvers used to get information out of quantum field theory. Nor do students need to be reminded of the many false starts and dead ends one finds in the history of science. For pedagogical reasons, then, conflicts and failures are largely ignored, leaving the impression that science pretty much has all the big problems wrapped up. With so few students getting beyond introductory courses, it's easy to see why this impression drives conventional wisdom. But it is this impression that paves the road toward naive realism. It places far too much faith in our ability to fully understand a highly complex world.

The second important idea is that theories and law statements[7] often break down outside of their particular domains (Section 3.4.2). Some examples are simple, for example, FLRW models of cosmology are not applicable close to the Big Bang.[8] The equations "blow up," throwing parameter values into impossible ranges. Other examples are more subtle. Even within classical mechanics itself, one must shift from one set of law statements to another depending on the scale and expected behavior of the system in question (Wilson 2009). Ordinary differential equations can capture the dynamics of point particles but only so long as collisions are ruled out. Since particles do, of course, collide in many systems, one must then replace point particles with rigid bodies. That is, unless elasticity becomes important, then one must use the deformable bodies found in classical continuum mechanics. Here, the mathematics becomes far more complicated; the existence and uniqueness of solutions to the governing equations are no longer guaranteed. (In fact, the only way to ensure that determinism holds across the whole of classical mechanics is to *assume* it as an axiom, as V.I. Arnold (1997) does in his.) The point, again, is that there is no one set of laws or a univocal definition of 'particle' that applies across every scale in classical mechanics. As Larry Sklar points out, things are no different in cutting-edge physics:

> It may turn out to be the case … that fundamental physical theory will consist of an infinite hierarchy of theories, each dealing with its own limited domain, but all linked together by their place in the hierarchy. And, in addition to this, it may very well turn out to be the case that no one of the partial theories in the hierarchy can be properly understood in its internal features without making reference to its place in the overall structure. (2002, 134)

The third idea needed for a more accurate scientific realism was dealt with in the previous chapter. The grand unification of all scientific knowledge—often under the banner of reductionism—does not describe current science or the science of the foreseeable future. The sciences are fragmented, highly specialized, and don't seem to be suffering much for it. The disconnectedness of the sciences is not an aberration. This is the normal state of things.

Scientific realists, in contrast, often treat science as if it were a single, massive body of knowledge. "*Science* aims at providing a true description of the world. *Science* is too successful to be false." It's as if science were a train advancing along a single track toward a better understanding of reality. Antirealists also tend to treat science as a whole, only to deny what realism affirms. In my view, there is something right about scientific realism, but we must also take account of the fragmentation and lack of coherence across and within different fields. *Science* does not make progress. Advances come piecemeal, as Mark Wilson argues:

> [I]f we no longer demand that science advance in great blocks of coherent framework built upon well articulated hunks of theory-to-measurement presupposition, we can better respect the fact that real life science only "comes at us in sections".... (2010, 566)

In short, scientific realism should be understood locally, not globally. Science does not advance along a single axis, like a train on a track. A better metaphor would be the many ways of approaching an island—boat, helicopter, submarine, hot air balloon, etc.—none of which are coordinated with each other. Theories within each field and subdiscipline can be approximately true but advancing along their own axes. While there is only one physical reality, the many independent paths for understanding and explaining that reality might never fit together in one neat package. The idea that we should be able to put the whole of science together in a perfectly coherent form is driven by overconfidence in our own abilities and a failure to acknowledge the tensions among different areas of knowledge.

With all that, the no-miracles argument still provides reason to believe that many of our best theories are in the right neighborhood, even if they are not true *simpliciter*. The independent advances of different theories, including engineering applications and unexpected new discoveries, give scientific realism—or at least the *local realism* advocated here—some justification. That this justification is defeasible (i.e., a given theory might turn out to be wrong after all) is not itself a problem. Virtually everything that we believe is defeasible. The fact that we *could* be wrong, or that we have been wrong in the past, should not push us into antirealism or skepticism. It's possible that what we think of as reality is artificially constructed, like a *Matrix* world. Given this possibility, should we therefore believe that our senses do not provide more or less accurate information about our environment? No, the proper response is to trust our senses for the most part, all the while knowing that they are fallible. Fallibility doesn't require antirealism about one's perceptions, reason, or memory. Antirealism, like skepticism more generally, is an overreaction to our imperfect epistemic access to a messy world.

The realism/antirealism debate is a perennial one in the philosophy of science, and I don't pretend to have given the definitive answer on it. The view presented here is, however, well within the mainstream among philosophers of science today.

7.4 Realism and Religion

So what does all this mean for religion? Here, I want to buck the common assumption that science and religion should be understood in fundamentally different terms: science about reason and observation and religion about faith and value. Only since the nineteenth century would anyone have suggested that religious belief is merely a matter of faith, where 'faith' is understood as a wholly different epistemological category from reason, one that has nothing to do with warrant or justification. Instead, both science and religion should be understood in realist terms. Both are attempting to provide at least an approximately true understanding of reality, albeit different aspects of reality for the most part.

Many students recoil at this idea. They and their professors like to think of themselves as empiricists—believing in what they can observe and being at least skeptical of all else. But as we've seen, one cannot be a scientific realist and a strict empiricist. God is unobservable but so are electrons, quantum mechanical entities in superposition, and the inside of black holes. Realists in both science and religion hold that one can have good reasons for believing in the existence of entities that cannot be directly observed. Granted, there is more evidence for unobservable entities in physics than in religion, but this is a difference in degree. The point is that the inability to observe God does not bar theism from being interpreted in a realistic manner. When it comes to entities that are beyond the reach of our senses, the question turns to what justification there is for believing that they exist. We'll consider that a bit more in Section 7.6.

Scientific realism is usually coupled with approximate truth. Without some notion that allows for relative degrees of closeness to reality, we would be forced to say that, for example, Newtonian physics is simply false. Mechanical engineers would therefore be spending most of their under-graduate studies on a false theory. NASA scientists would be using a false theory in order to predict the orbits of satellites. Scientific realists believe instead that more nuance is needed when describing successful theories that have been superseded. Approximate truth, as we've seen, is one way of describing the relative "rightness" of theories and laws.

Approximate truth can also be useful in understanding the relation between religions. First, let's get past the sophomoric claim that says all religions teach pretty much the same things. They don't. Saying that they do is usually a sign of ignorance or wishful thinking. There are, however, varying degrees of overlap between the teachings of many religions. Several faiths agree on monotheism, for example. This means that Islam is not simply false from a Christian point of view.[9] Islam, Judaism, and Christianity agree on a number of theological and ethical doctrines. So while a Christian theist would believe that his/her views are closest to the truth, other monotheistic faiths should be understood as approximately true, just less so.[10] Of course, Muslims would arrange the order differently, and that's fine. This is no different than any unresolved controversy in science. String theorists in physics disagree with advocates of loop quantum gravity. One of these approaches is, presumably, closer to the right answer than the other even though we can't say definitely which one it is. Likewise, there is one religion that is closest to the truth even if we cannot prove which. (If metaphysical naturalism is correct, then religions are mostly false.)

The point is that approximate truth provides a way of understanding the relation between religions without downplaying their differences. The major religions can be viewed in realist terms, as having varying degrees of truth. Unlike science, there will be less chance of resolving the differences or arriving at much consensus about which religion is most true, at least on this side of the afterlife. Many think this is deeply prob-lematic for theology, leading to skepticism or relativism. In my view, if science suffers from inconsistencies, tensions, and a general inability to fit together our best theories, we shouldn't expect anything more from religion. Science has the benefit of hands-on testing and direct feedback from nature. The *data* on which theology is based is comparatively much leaner, undermining the convergence of theological knowledge. That's frustrating to some degree, but frustration at a lack of complete understanding is part of every field. As the Beatles aptly put it, "We all doin' what we can" ("Revolution," 1968).

7.5   Models

Let's consider a closely related topic in the philosophy of science that might also prove useful in religion: models.

7.5.1   Models and Science

Roughly, a model is a representation of some object, behavior, or system that is usually not considered fully realistic. One familiar example is the *physical model*: a material, pictorial, or analogical representation of an actual system. Some are scale models, like the model planes used in wind tunnels. Scale models are useful when the laws governing the subject of the model are either unknown or too computationally complex to derive predictions. More common are *simplifying models* that abstract away properties and relations from what is being represented. Here, we find the usual zoo of physical idealizations: frictionless planes, perfectly elastic bodies, point masses, etc. These devices are "useful fictions" designed to simplify the mathematics. They are more or less realistic depending on the type of system being modeled and the amount of idealization involved.

Simplifying models are customized for different purposes. When designing an electric circuit, engineers often ignore nonlinearities by treating the system as if it were linear. If the question is under what conditions the circuit will become unstable, leaving the nonlinear components in place may be essential. The ability to pick and choose the kinds and degrees of idealization employed allows simplifying models to play many useful roles, from explanation to state prediction and stability analysis.

Sometimes, models are built in a "bottom-up" fashion in the sense that there are no first principles or laws of nature from which to derive them. These are called *phenomenological models*. For example, with enough data, one can build a computer model that simulates the behavior of city traffic, even though there are no general equations governing traffic flow. Simplifying and phenomenological models are just two examples of a rich typology.[11]

Some useful models are incompatible with each other. In nuclear physics, the nucleus is described using both the liquid drop model and the shell model (Morrison 2011, 346–351). The former is based, unsurprisingly, on an analogy with the behavior of a drop of fluid. The latter is a more bottom-up approach starting with the properties of protons and neutrons. While both models are used extensively, they describe contradictory mechanisms for the internal dynamics of a nucleus. The target system cannot have the properties ascribed by both. The exact right answer is beyond either of the models, and there isn't any firm basis for knowing which of the two is more realistic. Both models are useful in their own way, and it may be that "useful" is the best we can achieve given the limited epistemic access we have to some systems.

7.5.2   Models and Theology

My suggestion is that some theological disputes can best be understood as a conflict of models rather than one of absolute doctrines. Although theologians generally don't use the word 'model,' the idea of inexact representations has long been important in theology:

> In the Christian tradition, we use personal language about God, not because we think God is an old man with a beard sitting high above the bright blue sky, but because it is less misleading in using the finite resources of human language to call God 'Father' than it would be to employ the impersonal language of 'Force.' (Polkinghorne 2007, 34)

The anthropomorphic language used in the Hebrew Bible, such as talk about God's right hand or being seated on a throne, has traditionally been understood as metaphor. Theological doctrines, on the other hand, are not approached in this manner. Take for example the so-called "theories of atonement" in Christianity. What exactly was Christ's death on a cross supposed to accomplish, theologically speaking? There have been a number of theories put forward over the centuries with names such as penal substitution, ransom, recapitulation, Christus Victor, and others. The debate has typically been over which of these is the right theory, understood in naively realist terms: there is only one right answer; the different theories cannot be harmonized; as a matter of logic, they can't all be right.

A better approach would be to see many and perhaps all of these "theories" as incomplete, complimentary models of a complex theological truth. The exact right answer is inaccessible and perhaps incomprehensible to us. Instead of trying to debate the merits of each proposal in the search for the one true theory of the atonement, Christians can accept them all as different models of one event.[12] They are all useful as ways of understanding a complex and multifaceted truth. Polkinghorne has likewise argued that when it comes to theology,

> one may have to settle for a portfolio of different models—none claiming an exhaustive correspondence with the ways things are, but each usable with discretion—in order to cast light upon an appropriate range of phenomena.
>
> I do not think we should be discouraged by the realistic modesty of the remarks. Even in science, when we move away from the comparative simplicities of elementary particle physics into the complexities of condensed matter theory or the biology of organisms, we have to settle for whatever intellectual gains we can get. (1994, 36)

Theological models would function more like phenomenological models than simplifying ones. One resorts to phenomenological models, recall, when an overarching theory is lacking. As Polkinghorne notes, a portfolio of useful models is often the best one can do in theology. Useful for what? Matters of religious practice, hope, praise, prayer, or character development (e.g., actually becoming a more forgiving person rather than merely holding up forgiveness as an ideal). There is much in the fabric of religion that goes beyond a set of doctrinal beliefs.

Theists acknowledge our limited epistemic access to a supernatural reality. That every religion has some things wrong in it is not a recent discovery. Nonetheless, the inability of theology to get at the one perfect answer, solve all the questions, etc., is not unique. In the end, it's just part of being human. Science, philosophy, and theology are in the same boat. The suggestion here is that an understanding of how models are used in science might be useful in thinking about incomplete knowledge in religion. Instead of fretting or giving up, science uses models to help bridge the gap. Theology could do more of the same without sliding into relativism.

I close this chapter with a discussion about what is supposed to be the fundamental divide between religion and science: the former is based on faith, while the latter is based on reason.

7.6   Faith, Reason, and Trust

That science is based on "reason" is a vague platitude. What specific principles of reason does science use? The answer is not as straightforward as one might think. The history of science shows that many different forms of inference have taken center stage.[13] The scholastic medievals followed Aristotle. They believed that the key to understanding was the *apprehension* of a thing's essence. One first examines a large number of particulars, using the senses to investigate the samples. The next step is to abstract the universals they have in common. For example, the unique set of universals instantiated in every triangle constitutes the essence of a triangle. The method is the same for apprehending the nature of cats, stones, and every other natural being. Once a mind has grasped an essence, deduction is used to draw inferences. Much of medieval science involved deductive inferences from previously discovered essences.

Francis Bacon (1561–1626) was a powerful critic of this Aristotelian method. He believed that the medievals had overemphasized deduction and that induction should be the center of scientific knowledge. Descartes (1596–1650) went in the opposite direction, believing that axiomatic geometry provided the ideal: begin with "clear and distinct" ideas around which to form first principles and then apply deductive logic as far as it will go. Newton (1642–1727) thought that Descartes was far too optimistic in thinking that intuition alone could yield first principles. His method of analysis and synthesis was similar to the Aristotelian induction–deduction but without the metaphysics of essences. David Hume (1711–1776) and Immanuel Kant (1724–1804) later revealed deep conceptual problems with induction as a way to understand unobservable reality. Their arguments led to the radical empiricism of philosopher–scientist Ernst Mach (1838–1916), who in turn was strongly influential on the young Einstein. In the twentieth century, Karl Popper argued that how one arrives at a hypothesis—whether induction, intuition, or flashes of insight—is unimportant. It is instead the ability to falsify claims that is the key to science. Our best theories are merely the ones that have not yet been falsified, which is still an influential idea among scientists.

All this shows that while science surely relies on "reason," what that means has changed over time. Considering both the history of science and the sciences today, most philosophers now conclude that there is no such a thing as *the* scientific method. From the history of mechanics alone, what "one sees is a perpetual rethinking of a number of very fundamental ideas about just what scientific method ought to be and what the contents of successful science 'must' look like" (Sklar 2012, 87–88).

The broadest principle of logic employed by all the sciences is known as *abduction* or *IBE*: among the reasonable explanations for some evidence, the best explanation is closest to the truth, where truth is usually understood in a realist sense. Consider an auto mechanic. A customer brings in a car with one or more problems. The mechanic drives the car himself and then puts it through some diagnostic tests. He then forms the best explanation for all the information in hand and acts on that explanation, believing it to be at least approximately true. If that doesn't sound terribly "scientific," that's because we all use IBE, scientists or not.

So what makes one explanation better than another? This is where metatheoretic shaping principles (Section 1.3.1) come into play, especially the so-called explanatory virtues. Rival explanations are compared with respect to empirical accuracy, internal and external consistency, predictive success, simplicity, clarity, scope, and fruitfulness in guiding future research. The best explanation generally has several of these in its favor. Controversies arise when one explanation has, say, simplicity and clarity on its side, while another provides better predictions. This was the case for the earliest Copernican models. While far simpler than the older Ptolemaic calculations, they were not quite as accurate. Both sides claimed to have the better explanation for the data.

While IBE is ubiquitous, it is not without controversy.[14] The main problem is how to connect the explanatory virtues to truth. We no doubt prefer simple explanations to complex ones. That's why most investigators prefer the view that Lee Harvey Oswald acted alone in the assassination of President Kennedy to any of the many conspiracy theories. The latter are unnecessarily complex. But why, ask the critics of IBE, should one treat *our* preference for simplicity as an indicator of truth? How are truth and simplicity connected? The same question goes for clarity, scope, and the rest. Perhaps, these preferences are merely aesthetic.

The main response to this challenge is pragmatic. The explanatory virtues, like all shaping principles, have proved themselves useful over time. These principles have generally pointed the way to good theories, and good theories, says the scientific realist, are taken as approximately true. It would be nice if there were a necessary connection between metatheoretic shaping principles and truth, but there is none. There are no guarantees. We can either retreat into skepticism or press on with the best metaphysical and epistemic views available, none of which are written in stone.

Work on IBE in the philosophy of science has caught the attention of others, especially in analytical metaphysics and the philosophy of religion. If science can infer the existence of unobservable entities and their properties, the theory goes, so might philosophy. Metaphysicians therefore use IBE in making their case for the nature of properties and substances. Many of the arguments for the existence of God weigh naturalistic explanations of some phenomena against theistic ones. The list goes on. The point is that philosophers take IBE to be just as useful as scientists when working with rival explanations. Theologians and bible scholars do the same, although their "data" include scripture, the interpretations of scripture by those closer to the source (e.g., church fathers), and new information from history, classical studies, and archeology.

One objection to using IBE in philosophy and theology is that, unlike science, explanations in these disciplines lack significant "push back" from reality. Scientific explanations must survive empirical testing and the scrutiny of the scientific community. No matter how good an idea might be in the eyes of some researcher, "in physics, nature soon gets its turn" (Van Fraassen 2002, 29), and reality doesn't much care about our opinions. In time, new data is collected and experiments might fail to confirm a proposed explanation. And while falsification isn't as clear cut as Popper once thought, it is usually possible over time to disconfirm a hypothesis to the satisfaction of the scientific community. Not so with metaphysics or religion. Those explanations cannot be tested in any robust sense. Moreover, the criteria for "best explanation" in these disciplines shift according to current trends. A good explanation today would have been considered meaningless in the eyes of the logical empiricists less than a century ago.

In reply to this worry, we should agree that most scientific claims can be tested in ways that philosophy cannot, but that alone ought not prevent metaphysics and the philosophy of religion from using IBE. Every discipline needs investigative tools in order to push forward, and this one is a much too common principle of reason to ignore. But there should be caution. It is too easy for debates in these areas to become free floating, detached from any basis in reality. Seemingly important issues degenerate into philosophical constructions "whose appeal lies mainly in the logical problems they engender and the virtuoso displays of ingenuity we can then enjoy" (Van Fraassen 2002, 30). A proliferation of published articles does not entail that knowledge is accumulating.

Philosophers and theologians therefore need to work hard at keeping the ivory tower connected to the ground.

This worry about philosophical views becoming detached from reality applies to theology as well. Such concerns were responsible in part, I believe, for the Protestant emphasis on *sola scriptura*. When scripture is treated as data, it provides push back to creative innovations. As theological notions become detached from scripture and experience, there is a greater possibility of error and of old errors propagating over time.

I've been arguing in this section that science, philosophy, and theology are all based on reason, at least in part. In fact, they use the same broad principle of reason, IBE. But what about faith? Conventional wisdom says that faith is uniquely part of religion. Scientists, we're told, eschew faith and demand evidence. Even if both science and religion use IBE, only the latter is based on faith.

Perhaps. Whether this view is accurate depends a great deal on the notion of faith employed. Some understand faith as an act of will: one must choose to believe "by faith" when evidence is lacking. In a similar vein, Kierkegaard endorsed a "leap of faith" as something one does, an act initiated by the will rather than reason. A different approach takes faith to be its own cognitive capacity distinct from reason and emotion. Just as some people are more analytical or creative than others, on this view, some have a capacity for faith that allows them to accept religious truth, while others cannot. This is not a matter of will, but ability. Some have the "eyes of faith," while others don't. Faith in this latter sense can grow by exercising and strengthening this cognitive capacity over time.

The ancient Greeks had a rather different understanding of faith. The word we translate as 'faith' (*pistis*) was not a religious notion. *Pistis* was roughly synonymous with trust. One might have faith in the city walls or faith in a person to keep his or her word. A good king, like some of the gods,[15] could be trusted to hold immoral people accountable for their behavior. Jews and Christians adopted this broad idea and used it to describe their trust in God. The point is that *pistis* was a common idea that was sometimes used in a religious context. There was nothing mystical about it. This, I believe, is the primary notion of faith found in the New Testament and Septuagint. The other views of faith mentioned in the last paragraph evolved later.

Understood this way, science is as much about faith as any other aspect of life. For modern science to work, we must generally have faith/trust in our senses, reason, the testimony of others about their research, and the peer review process. We also have to have faith that the laws of nature will not radically change in the coming days or, more broadly, that past experience will continue to be a reliable guide for the future. Only an omniscient being can get by without faith. The point is that if faith is understood as *pistis*, then it ceases to be mystical or inherently religious. Scientists need faith just as much as anyone else.

In the end, there is nothing unique about the use of reason in science. Physicians, computer programmers, auto mechanics, and, yes, even philosophers and bible scholars all use IBE. Moreover, we all have to have faith in limited resources and in the knowledge passed on by others. What makes natural science unique is its access to new empirical data. In that, it is unrivaled by any discipline. That advantage, however, is merely a matter of resources. It is not a basis for segregating science and religion into the realm of reason on one hand and the realm of irrationality on the other.

7.7   Anomalies and Mystery

Let's finally consider how science and theology deal with information that conflicts with the prevailing view and with our limited epistemic access to reality.

While advances in science move at a faster pace than in the philosophy of religion and theology, progress has been made on a number of difficult issues. Consider the problem of evil. Unlike a century ago, scholars on all sides now agree that it is logically possible for God to be omniscient, omnipotent, and omnibenevolent and yet have a world with murder, disease, and the like. One need not reject classical theism in order to accommodate the presence of evil.

There is, however, the persistent feeling that the problem of evil has not been fully solved. It isn't clear, for example, why an innocent child of good and loving parents dies of cancer. Theists believe that God has a reason for allowing it, and that reason might be found among the many answers and theodicies in the literature. Sometimes, however, the answer is beyond our reach and perhaps requires a God's-eye perspective. In any case, the amount and kinds of evil in the world are not what one would expect from an omnibenevolent being. Orthodox theology leads to expectations that conflict with experience—thus the problem.

What is less appreciated is a similar dynamic in science. Theories lead to expectations about how the world should look, but things don't always turn out that way. Phenomena that are contrary to expectations are *anomalies*. Around the turn of the last century, physicists were confronted with hard anomalies like the photoelectric effect and blackbody radiation. I say "hard" anomaly to emphasize that these were not merely difficulties for classical atomic theory; they were logically inconsistent with that theory. Atomic models at the time *could not* be correct in light of this new data. Hard anomalies force a change. In this case, the change was from classical to quantum mechanics. Soft anomalies, on the other hand, are unexpected but not strictly inconsistent with a theory. While the irreducibly complex biological systems discussed earlier (Section 5.2) are problematic from a Darwinian point of view, they are considered soft anomalies even by most proponents of intelligent design. This means that while irreducible complexity constitutes a problem to be solved, it is logically consistent with neo-Darwinism as it stands. Soft anomalies do not force a change in the governing theory, although they do motivate researchers to look for other options.

In the face of anomalies, progress tends to be conservative (Section 1.3.1). In other words, scientific theories will and should change as little as needed in order to accommodate new discoveries. Evolving theories are the norm; revolutions should be the exception. Consider the history of quantum mechanics. Classical physics was not discarded in light of anomalies such as blackbody radiation. Instead, a handful of quantum principles were added to classical models of the atom long before the quantum mechanical revolution in the mid-1920s. The hope was that the classical picture of the atom could be preserved with only a few new constraints. The hybrid classical/quantum approach, now known as the "old quantum theory," worked well for the hydrogen atom but failed for helium. Conservative changes were not sufficient in this case. This episode also illustrates the shaping principle known as *tenacity*: good theories earn the right of continued acceptance even in the face of some anomalies. The reign of classical mechanics did not end quickly. Only after all attempts to revise more conventional models failed did a full-blown quantum mechanics emerge.

The lesson here is that both scientific theories and theological doctrines have anomalies. Moreover, both disciplines react similarly: well-established views tend to be tenacious and revisions tend to be conservative. This frustrates advocates of change, but tenacity and conservativeness serve useful roles. They keep intellectual fads from quickly dominating the scene and force new ideas to prove themselves before becoming mainstream.

What about those cases where anomalies persist and questions remain unanswered? Many think that the problem of evil is in this category as well as paradoxes surrounding the incarnation and Trinity. Philosophers and theologians have made progress on these questions, but no one seems to have a completely satisfying answer. At least scientists don't have to worry about intrinsically unresolvable problems like these. Right?

Well, "unresolvable" is a strong term. No one can say what discoveries might be made. But for now, there do seem to be limits to a scientific understanding of nature. Several persistent questions have to do with how quantum mechanics applies to the macroscopic world. These include how to solve the measurement problem and how classical chaos emerges from a quantum mechanical base.[16]

The situation is not quite as bad in cosmology, but there are problems. While general relativity predicts that the universe began at the Big Bang, the Big Bang itself is a singularity; some values in the equations "blow up" (i.e., become infinite).[17] In this case, the infinities involve density and spacetime curvature. Saying that relativity "predicts" a Big Bang is not strictly true. Once a set of parameters blow up, the equations no longer apply. What we call the Big Bang is precisely where general relativity no longer makes physical sense.[18] While progress has been made and a number of ways to get around the singularity have been proposed, there is no consensus. Given the limited epistemic access we have to that time, I doubt there will ever been a widely accepted alternative. We will never be able to reproduce the extreme conditions at the Big Bang, even though particle accelerators have been getting closer. In the end, the correct answer will likely be beyond our reach, as Polkinghorne rightly says: "History suggests that quite severe limits should be set on any expectation of human ability to second guess nature in regimes lying far beyond current experimental access" (2007, 27). The bottom line is that while scientific knowledge will continue to expand, there are limits. That shouldn't be surprising. We are finite beings with fixed resources. Some corners of nature will remain at least partially shrouded in mystery.

7.7.1   Which Brings us Back to Theology

There, *mystery* is a technical term. It refers to truths that, if we are to know them, must be explicitly revealed by God. The New Testament describes the full understanding of Jesus as Christ, for example, as a mystery (*mysterion*) that has been revealed to the church (Eph 1:8–10, Col 2:2–3). The appeal to mystery is made when there is an unbridgeable epistemic gap between us and the supernatural. Unlike scientists, theists are often criticized for this move. "It's a cop out," says the atheist, "the reason you can't answer the question is that there is no good answer." This response is understandable.

If the appeal to mystery is the best the theist can do in the face of difficult questions, then theism would seem to be an empty husk, dying "the death by a thousand qualifications" (Flew 1955).

My suggestion is that a better understanding of anomalies, singularities, and the limits of scientific knowledge could add some useful perspective to the notion of mystery in theology.

Consider again the idea of the Trinity. Christians claim to be monotheists who believe that God exists in three persons. If there are three persons—three centers of consciousness—then there would seem to be three gods. If monotheism is true, on the other hand, then by definition there can only be one God. How then can one be a trinitarian monotheist?

Recall the distinction between the laws of nature and law statements (Section 4.3.2). The laws of nature are the universal regularities that scientists are trying to discover. Law statements are the equations and propositions found in textbooks. The former are "out there" in reality, in some sense or other. The latter are what we believe the laws to be. There is a similar distinction one can make here. Singularities are often understood as artifacts of our models rather than entities or events in nature itself. When it comes to cosmology, there was a way the universe actually began (nature), and then there are the relativistic models used to mathematically describe that beginning (model). Many physicists, following Einstein, believe that while the equations say that the density of the universe was infinite at the Big Bang, that is precisely where the models break down (artifact). The singularity is due to the limited nature of the models themselves. That doesn't mean that the models should be rejected; rather, one must be mindful of their domain of applicability.

7.7.2   Back to Theology

The Trinity might best be understood as something like a singularity of monotheism and personhood. Both are useful concepts, but they are generally applied in separate contexts. Monotheism is an intrinsically religious idea, but personhood crops up in more diverse ways. For example, the latter is used in matters as different as the abortion debate (is the fetus a person?) to thinking about extraterrestrials (if there were such beings as Klingons, they would be nonhuman persons). These two concepts are pushed to the extreme when talking about three persons being one God. The reason that there is no one accepted model of the Trinity is that the categories of monotheism and multiple persons get stretched beyond their limits. Whatever is actually the case (reality) cannot be accurately captured in terms that we fully understand (model). That doesn't mean that God cannot be a Trinity any more than the universe cannot have started with a Big Bang. A singularity does not refute the laws or models in which it appears. Instead, the concepts of monotheism and multiple persons, which work well on their own, cannot be extended into the theological singularity itself.[19]

The appeal to mystery/singularity is not a blank check in either science or theology. It is a card, to change the metaphor, that must be played judiciously. It ought not be used to simply protect a bad theory or doctrine. Nonetheless, one of my take-aways as a philosopher of science is that while we would like to have all of the answers in cosmology, particle physics, and everything in between, we often have to settle for something much less. The philosophy of science helps one come to grips with why it is that we don't— and perhaps can't—understand everything about nature. Our epistemic resources eventually run out. But if there are limits to knowledge in science with its endless data, controlled experiments, and hands-on access to the systems being investigated, how could one not expect the same in matters involving metaphysics and the supernatural?

One might ask, as Dale Tuggy has (private correspondence), if the appeal to mystery is not a blank check, under what conditions can it be used? I think there can be no formulaic answer. The decision will involve trade-offs between theological shaping principles, in this case precision and internal coherence versus conservatism and the claims of scripture. One would like clear and precise theological doctrines that involve no contradictions or paradoxes.[20] One would also like doctrines that fit with scripture and involve no change to core dogma. Unfortunately, you can't always get what you want (Rolling Stones, 1969). (Okay, no more lyrics.) On one hand, the doctrine of the Trinity has not been precisely articulated and is *prima facie* contradictory. On the other hand, it maintains continuity with its monotheistic Jewish roots and is the best reconstruction of biblical claims regarding God the Father, Jesus, and the Holy Spirit. Trinitarians take the latter *desiderata* to trump the former. Others have rejected this part of orthodoxy in favor of other options. There is no theological meta-rule that can decisively resolve this, much as the scientific method is useless in resolving tensions between metatheoretic shaping principles.

In the end, we see that both science and religion involve matters of faith and knowledge, theory and model, and success and anomaly. This is not surprising if we think of the two in realist terms—complementary disciplines that usually try to make sense of different aspects of reality (Section 1.4.6). Every truth-seeking endeavor will have some combination of data, explanation, educated posit, progress, and failure, although each discipline has its own technical terms to describe them. The goal of this chapter was to present some of the tools developed in the philosophy of science that might prove useful in religion and theology. This is not the imposition of one field's jargon on another. It is rather the offering of conceptual hooks that experts in other areas might find helpful. If not, go in peace. Interdisciplinary work always looks better on paper than it does once scholars get put in a room together. For my part, the philosophy of science has been helpful in understanding controversies far removed from physics. Perhaps, others will benefit as well.

Notes

1   "When I think about the ablest students whom I have encountered in my teaching, that is, those who distinguish themselves by their independence of judgment and not merely their quick-wittedness, I can affirm that they had a vigorous interest in epistemology" (Einstein 1916, 101).

2   That's where the idea of using leeches in order to draw off excess blood comes from. On the humoral theory, people display symptoms of diseases when their four humors are out of balance. It was replaced with more modern views of contagion in the nineteenth century.

3   I am intentionally glossing over subtle issues regarding truth-bearers and truth-makers. See Rodriguez-Pereyra (2006) for more.

4   While Kuhn's work was the most influential across all fields, Rodney Holder rightly suggests (private correspondence) that Karl Popper's name is best known to scientists.

5   This is sometimes called the *postmodern* view of science. By whatever name, this category is extremely nebulous. See Kukla (2000) for more.

6   In the 1999 film *The Matrix*, humans unknowingly exist in a computer-generated reality. Hence, almost every belief held within the *Matrix* world is false.

7   The distinction between laws and law statements was discussed in Section 4.3.2.

8   Prior to Planck time, 10^{-43} s after the Big Bang.

9   Granted, the bald assertion "Islam is a false religion" is more likely to be heard in the context of a sermon than in a scholarly text. My point is that no Christian with a basic understanding of Islam literally believes that the latter is false *simpliciter*. Dale Tuggy has suggested (private correspondence) that a simpler approach would be to say that Christianity has more true claims than Islam, from a Christian point of view. The degree to which a religion is true is merely a matter of how many true claims it contains, perhaps with some penalty assessed for false claims. Such a view presupposes, however, that the content of either a religion or a scientific theory can adequately be reduced to a list of propositions. I doubt that, although spelling out that doubt takes us into difficult issues in the philosophy of language. See Wilson (2006, especially chap. 5).

10   This is a wholly separate matter from whether Christianity provides the unique means of salvation. Christian inclusivists and exclusivists can affirm that (i) Jesus of Nazareth provides the sole means of salvation, (ii) Christianity contains more truth than any other religion, and (iii) other religions are more or less approximately true.

11   For more, see Frigg (2008) and Koperski (2006).

12   Although I don't want to go into the details of the different views, it seems that they each correspond to different attributes or goals commonly ascribed to the theistic God: justice (penal substitution), compassion (ransom), changing humanity's fundamental course (recapitulation), and the defeat of evil (Christus Victor).

13   For an introduction and historical overview, see Losee (2001).

14   See Douven (2011) for a good introduction and van Fraassen (1989, 131–150) for more in-depth criticism. In a different vein, Bayesians will demand that all this loose talk of explanatory virtues be cashed out in probabilistic terms (McGrew 2003). I wish them well, but do not believe that IBE stands or falls depending on their results.

15   The *Odyssey* opens with Athena and Zeus discussing the immoral behavior of Penelope's suitors. The lesson seems to be that unrighteous acts are noticed by the gods and will be reckoned accordingly (Lorna Holmes, private correspondence).

16   See Albert (1994) for the former and Belot (2000) for the latter.

17   There are other types of singularities. See Earman (1995, chap. 2) for more.

18   Or at least that was Einstein's understanding of spacetime singularities such as the Big Bang. That still seems to be the standard view of things, although Earman argues for a more nuanced approach (1995, 11–21).

19   This is a version of what Dale Tuggy (2010) calls *negative mysterianism*.

20   Not quite everyone would like this. Some of Tuggy's *positive mysterians* celebrate the paradoxes and contradictions in theology.

References

Albert, David. 1994. *Quantum Mechanics and Experience.* Cambridge: Harvard University Press.

Arnold, V.I. 1997. *Mathematical Methods of Classical Mechanics.* Berlin: Springer.

Belot, Gordon. 2000. "Chaos and Fundamentalism." *Philosophy of Science* 67 (3): 465.

Douven, Igor. 2011. "Abduction." In *Stanford Encyclopedia of Philosophy*, edited by Edward N. Zalta. Stanford: Metaphysics Research Lab, Center for the Study of Language and Information, Stanford University. http://plato.stanford.edu/entries/abduction/. Accessed August 2, 2014.

Earman, John. 1995. *Bangs, Crunches, Whimpers, and Shrieks: Singularities and Acausalities in Relativistic Spacetimes.* New York: Oxford University Press.

Einstein, Albert. 1916. "Ernst Mach." *Physikalische Zeitschrift* 17: 101–104.

Flew, Antony. 1955. "Theology and Falsification." In *New Essays in Philosophical Theology*, edited by A. Flew and A.C. MacIntyre, 99–100. London: SCM Press.

Frigg, Roman. 2008. "Models in Science." In *Stanford Encyclopedia of Philosophy*, edited by Edward N. Zalta. Stanford: Metaphysics Research Lab, Center for the Study of Language and Information, Stanford University. http://plato.stanford. edu/entries/models-science/. Accessed September 12, 2013.

Giere, Ronald N. 1999. *Science Without Laws.* Science and Its Conceptual Foundations. Chicago: University of Chicago Press.

Hajek, Petr. 2010. "Fuzzy Logic." In *Stanford Encyclopedia of Philosophy*, edited by Edward N. Zalta. Stanford: Metaphysics Research Lab, Center for the Study of Language and Information, Stanford University. http://plato.stanford.edu/ entries/logic-fuzzy/. Accessed September 12, 2013.

Koperski, Jeffrey. 2006. "Models." In *Internet Encyclopedia of Philosophy*, edited by James Fieser and Bradley Dowden. http://www.iep.utm.edu/models/. Accessed June 5, 2007.

Kukla, André. 2000. *Social Constructivism and the Philosophy of Science.* Philosophical Issues in Science. New York: Routledge.

Losee, John. 2001. *A Historical Introduction to the Philosophy of Science.* New York: Oxford University Press.

McGrew, Timothy. 2003. "Confirmation, Heuristics, and Explanatory Reasoning." *British Journal for the Philosophy of Science* 54: 553–567.

Morrison, Margaret. 2011. "One Phenomenon, Many Models: Inconsistency and Complementarity." *Studies in History and Philosophy of Science* 42 (2): 342–351.

Polkinghorne, John. 1994. *The Faith of a Physicist.* Princeton: Princeton University Press.

Polkinghorne, John. 2007. *Quantum Physics and Theology.* New Haven: Yale University Press.

Rodriguez-Pereyra, Gonzalo. 2006. "Truthmakers." *Philosophy Compass* 1 (2): 186–200.

Rorty, Richard. 1987. "Science and Solidarity." In *The Rhetoric of the Human Sciences: Language and Argument in Scholarship and Public Affairs*, edited by J.S. Nelson, A. Megill, and D.N. MacCloskey, 38–52. Madison: University Wisconsin Press.

Sklar, Lawrence. 2002. *Theory and Truth: Philosophical Critique Within Foundational Science.* Oxford: Oxford University Press.

Sklar, Lawrence. 2012. *Philosophy and the Foundations of Dynamics.* Cambridge: Cambridge University Press.

Stanford, P. Kyle. 2003. "Pyrrhic Victories for Scientific Realism." *The Journal of Philosophy* 100 (11): 553–572.

Stanford, P. Kyle. 2006. *Exceeding Our Grasp: Science, History, and the Problem of Unconceived Alternatives*. New York: Oxford University Press.

Tuggy, Dale. 2010. "Trinity." In *Stanford Encyclopedia of Philosophy*, edited by Edward N. Zalta. Stanford: Metaphysics Research Lab, Center for the Study of Language and Information, Stanford University. http://plato.stanford.edu/entries/trinity/. August 2, 2014.

Van Fraassen, Bas C. 1989. *Laws and Symmetry*. Oxford: Clarendon Press.

Van Fraassen, Bas C. 2002. *The Empirical Stance*. New Haven: Yale University Press.

Wilson, Mark. 2006. *Wandering Significance: An Essay on Conceptual Behaviour*. Oxford: Oxford University Press.

Wilson, Mark. 2009. "Determinism and the Mystery of the Missing Physics." *British Journal for the Philosophy of Science* 60 (1): 173–193.

Wilson, Mark. 2010. "What Can Contemporary Philosophy Learn from Our 'Scientific Philosophy' Heritage." *Noûs* 44 (3): 545–570.

Index

The Physics of Theism: God, Physics, and the Philosophy of Science,
First Edition. Jeffrey Koperski.
© 2015 John Wiley & Sons, Ltd. Published 2015 by John Wiley & Sons, Ltd.